Ian Wood

Elliptic and Parabolic Problems in Non-Smooth Domains

Bibliografische Information Der Deutschen Bibliothek

Die Deutsche Bibliothek verzeichnet diese Publikation in der Deutschen
Nationalbibliografie; detaillierte bibliografische Daten sind im Internet über
http://dnb.ddb.de abrufbar.

ISBN 3-8325-1059-1

Dissertation
TU Darmstadt
D17

Logos Verlag Berlin
Comeniushof, Gubener Str. 47,
10243 Berlin
Tel.: +49 030 42 85 10 90
Fax: +49 030 42 85 10 92
INTERNET: http://www.logos-verlag.de

Contents

Introduction

Partial differential equations (PDEs) play a fundamental role in the modeling of natural phenomena such as heat flow, fluid dynamics, electro-magnetism, population dynamics and many, many more. Therefore, they are an important area of mathematics with applications in many other sciences. One interesting question in this field of research concerns the regularity or smoothness of solutions to elliptic and parabolic PDEs. Amongst other things, the regularity of the solution depends on the regularity of the domain in which the equation is considered. For sufficiently smooth domains, there is a very satisfying theory, at least for a large class of linear elliptic and parabolic PDEs.

The aim of this thesis is to investigate the regularity of solutions to PDEs in Lipschitz domains. These domains can have boundary singularities, i.e. corners or edges, and in this case the situation becomes more difficult. While some results that hold in smooth domains are still valid in Lipschitz domains, the singularity in the boundary can lead to the solutions being less regular than one might expect from the data. In [Dah79], Dahlberg constructs a bounded Lipschitz domain $\Omega \subseteq \mathbb{R}^2$ and a function $f \in L^\infty(\Omega)$ where the solution to

$$(1) \qquad \begin{cases} \Delta u = f & \text{in } \Omega, \\ u = 0 & \text{on } \partial\Omega \end{cases}$$

is not in $W^{2,p}(\Omega)$ for any $1 < p < \infty$. In Section 3.4, for $p > 3$, we will construct a bounded Lipschitz domain $\Omega \subseteq \mathbb{R}^n$, $n \geq 3$, where the solution isn't even in $W^{1,p}(\Omega)$ for some $f \in L^p(\Omega)$.

The study of elliptic problems in non-smooth domains was initiated in the 1960's with works by Agmon and Nirenberg [AN63]. In 1967, Pazy was able to give an asymptotic expansion to solutions in Hilbert spaces near corners [Paz67]. The main idea was developed by Kondratiev [Kon67] and has since been extended, in particular by Grubb [Gru86], Maz'ja, Nazarov and Plamenevsky [MP84] and [NP94] and Schulze [Sch91]. In this approach, the first step is to consider a model problem in a cone. By a transformation of coordinates, the cone can be mapped to a cylinder. Using partial Fourier transforms in the direction of the cylinder axis, one obtains an operator pencil and has to solve the corresponding lower dimensional problems. The inverse Fourier transform yields the solution in the cylinder in weighted spaces and provides asymptotics for the solution. More general domains can then be treated using localisation methods.

A different approach to the problem is to use potential theory and harmonic analysis methods to estimate the Green function or harmonic measure. This approach was used by Dahlberg [Dah77] and [Dah79], and Jerison and Kenig [JK82] and [JK95] for the Dirichlet-Laplacian in bounded Lipschitz domains. It allows to prove uniqueness of the solution to problem (1) in certain Sobolev spaces when the data f is in a negative Sobolev space. Adolfsson [Ado92] and [Ado93] and Fromm [Fro93] show that in bounded convex domains Ω, or, more generally, bounded Lipschitz domains satisfying a uniform outer ball condition, the solution still gains two degrees of regularity when the data is in $L^p(\Omega)$. For the Neumann-Laplacian the corresponding results were proved by Fabes, Mendez and Mitrea [FMM98] and, in the convex case, by Adolfsson and Jerison [AJ94]. See Section 2 for more details.

For the parabolic problem

$$\begin{cases} u_t - Au = f & \text{in } (0,T) \times \Omega, \\ \quad Bu = 0 & \text{on } (0,T) \times \partial\Omega \end{cases}$$

in smooth domains Ω with an elliptic operator A of order $2m$ and a boundary operator B there have been many developments in recent years. They rely in large part on developments in operator theory: the analysis of Banach-space valued functions, the functional calculus of sectorial operators and the property of maximal regularity (cf. Section 1.3) for the linear problem. In [DHP03], maximal regularity of solutions to boundary value problems of Agmon-Douglis-Nirenberg type was proved for C^{2m}-domains.

For parabolic problems in Lipschitz domains relatively little is known so far. There are some results using classical methods by Kondratiev and Kozlov [KMR01] and Nazarov and Plamenevskij [NP94]. In [Gri85], Grisvard obtains strong solutions to the heat equation in $L^2(Q)$ where $Q = \bigcup_{0<t<T}\{t\} \times \Omega_t$ is Lipschitz in $(0, T) \times \mathbb{R}^n$ and for each $t \in (0, T)$, Ω_t is an interval. Griepentrog [Gri99] also considers parabolic equations in Hilbert spaces. Schrohe and Schulze (cf. [SS99] and the references therein) use pseudo-differential operators to obtain results in weighted spaces. Following Dahlberg's approach to the elliptic problem, Fabes and Salsa [FS94], use caloric measure to solve

$$(2) \qquad \begin{cases} u_t - \Delta u = 0 & \text{in } (0, T) \times \Omega, \\ u(0) = 0 & \text{in } \Omega, \\ u = f & \text{on } (0, T) \times \partial\Omega \end{cases}$$

in Lipschitz domains Ω for data in some L^p-space on the boundary, $p \geq 2$. In [Bro89] and [Bro90], Brown uses potential theoretical arguments to show existence and uniqueness of solutions of (2) in certain function spaces and to gain estimates on the non-tangential maximal function of the gradient of the solution. However, none of these methods seem to allow to prove maximal regularity for the operators or to characterise the domain in the L^p-setting.

In [She95], Shen investigated elliptic systems with constant coefficients in bounded Lipschitz domains. He was able to prove resolvent and gradient estimates in $L^p(\Omega)$ for p in an interval around 2. These guarantee that the operator generates a bounded analytic semigroup and that the domain of the operator is contained in $W^{1,p}(\Omega)$.

In this thesis we will investigate elliptic and parabolic problems in Lipschitz domains, in particular we want to combine the known results for elliptic equations with results from operator theory to solve the parabolic problem and obtain maximal regularity for the solution.

The structure of the thesis is the following. In Chapter 1 we present some of the fundamental concepts used throughout this thesis. Much of this is standard but we state it here to clarify notation and to keep the thesis as self-contained as possible. We introduce the function spaces in which we will be working, the non-homogeneous and homogeneous Sobolev spaces and the Bessel potential spaces and present results on density, embeddings, extensions and boundary values of functions in these spaces. In a section on operator theory, we then introduce C_0-semigroups, H^∞-calculus and bounded imaginary powers for sectorial operators, as well as the concept of maximal regularity and Fourier multipliers.

In Chapter 2, we consider Laplace's equation

$$(3) \qquad \qquad \Delta u = f$$

in Lipschitz domains Ω with the data f in some Sobolev space and Dirichlet or Neumann boundary conditions. Section 2.1 presents an overview of the main results known for bounded domains. These are the results due to Jerison and Kenig (Dirichlet problem), Fabes, Mendez and Mitrea (Neumann problem), Adolfsson and Fromm (Dirichlet problem in convex domains), Adolfsson and Jerison (Neumann problem in convex domains) and Shen (resolvent problem) already mentioned above.

In Section 2.2, we generalise these results in the case of Dirichlet boundary conditions for unbounded Lipschitz domains. To do this, we approximate the unbounded domain Ω using bounded Lipschitz domains Ω_R. We will show that the solutions u_R of (3) in Ω_R converge to a solution u of (3) in Ω in the homogeneous Sobolev spaces.

Chapter 3 is the heart of this thesis. In this chapter, we consider the heat equation

$$\begin{cases} u_t - \Delta u = f & \text{in } \mathbb{R}_+ \times \Omega, \\ u(0) = 0 & \text{in } \Omega, \\ Bu = 0 & \text{on } \mathbb{R}_+ \times \partial\Omega, \end{cases}$$

where Bu is either the boundary operator $u|_{\partial\Omega}$ or $\frac{\partial u}{\partial N}|_{\partial\Omega}$.

In the first section we consider the Dirichlet-Laplacian with domain $D_1(\Delta) = \{u \in W_0^{1,p}(\Omega) : \Delta u \in L^p(\Omega)\}$ in bounded Lipschitz domains Ω and with domain $D_2(\Delta) = W^{2,p}(\Omega) \cap W_0^{1,p}(\Omega)$ in bounded convex domains. Using the results from the elliptic theory in Chapter 2, in both cases we are able to give a range for the exponent p such that the Dirichlet-Laplacian generates a C_0-semigroup in $L^p(\Omega)$ (Theorems 3.1.8 and 3.1.9). We prove that the semigroups we have gained coincide with the semigroups obtained via the form method (see e.g. [Ouh04]), however, our approach has the advantage of characterising the domain of the generator. We also show some properties of the generator and the generated semigroup which is contractive, analytic and positive.

The second section is devoted to the Neumann-Laplacian which we also consider with two different domains. We proceed as in the case of Dirichlet boundary conditions to prove similar generation results (Theorems 3.2.5 and 3.2.7) and properties of the generated semigroup.

The next step, carried out in Section 3.3, is to prove our main result, maximal regularity for the Laplacian in Lipschitz domains. Combining the results of the preceding two sections with results from operator theory in Section 1.3, we are able to show that the Laplacian with domain $D(\Delta) = \{u \in W^{1,p}(\Omega) : \Delta u \in L^p(\Omega),\ Bu = 0\}$ has maximal regularity in bounded Lipschitz domains in $L^p(\Omega)$ for $(3+\varepsilon)' < p < 3+\varepsilon$, for some $\varepsilon > 0$ depending on Ω (Theorems 3.3.1 and 3.3.3). By making the additional assumption that Ω is convex, in the case of Dirichlet boundary conditions the range of p for which we get this result can be extended to all $1 < p < \infty$. In addition, if $1 < p \leq 2$ and Ω is bounded and convex, the Laplacian with domain $D(\Delta) = \{u \in W^{2,p}(\Omega) : Bu = 0\}$ has maximal regularity (Theorems 3.3.2 and 3.3.4).

In the final part of Chapter 3, we use results on harmonic functions in the complement of slender cones to construct a bounded Lipschitz domain and an unbounded smooth domain Ω such that the correct domain of the Laplacian in $L^p(\Omega)$, $p > 3$, i.e. the domain of definition $D(\Delta)$, such that $(\Delta, D(\Delta))$ is a closed operator in $L^p(\Omega)$, is not even contained in $W^{1,p}(\Omega)$ (Corollary 3.4.6).

In Chapter 4, we extend the results on the Laplacian from the previous two chapters to more general elliptic operators A with L^∞-coefficients. We consider both the elliptic problem $Au = f$ and the parabolic problem $u_t - Au = f$ in $L^p(\Omega)$ for convex domains Ω with $p \in (p_0, 2]$.

As can be seen from a well-known example presented in Section 4.1, even in smooth domains the solution to $Au = f$ with Dirichlet boundary conditions need not be unique. However, if the coefficients of A satisfy a so-called Cordes condition (cf. condition (4.2), or, for parabolic equations (4.11)), the results valid for the Laplacian carry over to A via a fixed point argument and we get a precise description of the domain and maximal regularity estimates for operators with variable coefficients (Theorems 4.2.9 and 4.3.9).

For elliptic equations in two dimensions strong ellipticity of A is sufficient to guarantee that A satisfies the elliptic Cordes condition (Corollary 4.2.11). For parabolic equations, unfortunately, this does not seem to be the case and we need an extra assumption that the coefficients of the operator do not vary too much in time or space for the parabolic Cordes condition to be satisfied (Corollary 4.3.15).

Finally, in Chapter 5, we consider Ornstein-Uhlenbeck operators. These operators are first order perturbations of the Laplacian with, in general, unbounded coefficients. They appear for instance, when considering the heat equation in rotating domains. Using a transformation, we obtain an equation of the form $u_t = Au := \Delta u + Mx \cdot \nabla u$ in a fixed domain Ω where M is a constant real-valued matrix M.

The case $\Omega = \mathbb{R}^n$ has been investigated by many authors, in particular, we refer to the papers by Da Prato, Lunardi and Vespri [DPL95] and [LV96] and to the works by Metafune, Prüss, Rhandi and Schnaubelt [Met01], [MPRS02], [MPRS04] and [PRS04]. They show that in $\mathbb{R}^n$, the Ornstein-Uhlenbeck operator $A = \Delta + Mx \cdot \nabla$ generates a positive semigroup on all $L^p(\mathbb{R}^n)$. Due to the unbounded term Mx this is not a simple perturbation result and in fact it can be shown that the generated semigroup is not analytic. We state the main results for the $\mathbb{R}^n$-case in Section 5.1.

If D is a bounded Lipschitz domain, Ornstein-Uhlenbeck operators can be treated using standard perturbation methods and a gradient estimate on the solution to the resolvent problem $(\lambda - \Delta)u = f$. In this way, we see that the Ornstein-Uhlenbeck operator A with the domain $D(A) = \{u \in W_0^{1,p}(D) : \Delta u \in L^p(D)\}$ generates an analytic semigroup on $L^p(D)$ for p in an interval around 2 and we can give an estimate on the growth bound (Corollary 5.3.1). If the domain satisfies a uniform outer ball condition, the operator A with domain $D(A) = W^{2,p}(D) \cap W_0^{1,p}(D)$ is the generator of an analytic semigroup on $L^p(D)$ for $1 < p \leq 2$ (Corollary 5.5.2).

By a cut-off procedure used previously by Hishida [His99b], we then combine the results in $\mathbb{R}^n$ and the results in bounded domains to obtain a generator result for Ornstein-Uhlenbeck operators in exterior Lipschitz domains (Theorems 5.4.1 and 5.5.5). The regularity for solutions in bounded domains translates into the same regularity for solutions of the Ornstein-Uhlenbeck equation in exterior domains. In particular, in domains Ω satisfying a uniform outer ball condition the domain of the Ornstein-Uhlenbeck operator is contained in $W^{2,p}(\Omega)$, $1 < p \leq 2$. Furthermore, in these domains, we are able to show so-called L^p-L^q-smoothing estimates (Lemma 5.5.3 and Theorem 5.5.7).

Acknowledgements

At this point, it is a great pleasure for me to thank those who, in some way or other, contributed to the development of this thesis.

This thesis was conceived and written at the Darmstadt University of Technology and during a one year stay at the University of Chicago while I was supported by a stipend from the Studienstiftung des deutschen Volkes.

First, I would like to thank my advisor Prof. Matthias Hieber for the chance to work on these interesting problems, his guidance and many helpful suggestions. I am particularly grateful to him for having been given the opportunity to spend part of the time working on this dissertation abroad and his help in arranging this.

A very special thanks goes to Prof. Carlos Kenig for the chance to spend a year at the University of Chicago and for many fruitful discussions during that period and since which contributed enormously to this work. In addition, I would like to thank him for acting as co-referee.

Also, I would like to thank Michael Korey for initiating the contact to Professor Kenig and valuable advice on living and studying in Chicago.

I am very grateful to Prof. Jan Prüss and Prof. Reinhard Farwig for pointing out some errors in the first version of the thesis and to both of them and to Roland Schnaubelt and Mathias Wilke for assistance in correcting those mistakes.

Further thanks go to the Studienstiftung des deutschen Volkes, whose financial support enabled me to work on this project and to finance my stay in Chicago.

I wish to thank my colleagues Eva Dintelmann, Matthias Geißert, Robert Haller-Dintelmann, Horst Heck, André Noll, Jürgen Saal and Julian Wiedl in Darmstadt and Gautam Iyer in Chicago for many interesting discussions and the open atmosphere in our research group. In particular, I thank Eva, Robert and Matthias for proof-reading different parts of this manuscript.

Finally, I thank my parents for their support and encouragement throughout the time of my work on this thesis and for proof-reading it.

Chapter 1

Preliminaries

This chapter serves as an introduction to some of the basic concepts that will be used throughout the rest of this thesis. To be more precise, we describe the function spaces that we need and state some of their basic properties and then introduce some results from operator theory on C_0-semigroups, bounded imaginary powers (BIP) of operators, H^∞-calculus, maximal regularity and Fourier multipliers. We start with a section on notation.

1.1 Notation

Throughout this thesis, $\mathbb{R}_+ = (0, \infty)$ denotes the positive real numbers. X and Y denote Banach spaces with norms $\|\cdot\|_X$ and $\|\cdot\|_Y$, respectively. We denote the continuous linear mappings from X into Y by $\mathcal{L}(X, Y)$ and by $\mathcal{L}(X)$ the continuous linear mappings on X. X' denotes the dual space to X and $\langle x, x' \rangle$ the dual pairing for $x \in X$ and $x' \in X'$. $B_R(x)$ will denote the open ball of radius R centered at x. If $x = 0$, we simply write B_R. Furthermore, $\mathbb{S}^n := \{x \in \mathbb{R}^{n+1} : |x| = 1\}$ denotes the n-dimensional unit sphere.

δ_{ij} denotes the Kronecker delta, i.e. $\delta_{ij} = 1$ for $i = j$ and δ_{ij} is zero otherwise. With D_i and D_{ij} we denote the differential operators $\frac{\partial}{\partial x_i}$ and $\frac{\partial^2}{\partial x_i \partial x_j}$, respectively. For a multi index $\alpha \in \mathbb{N}_0^n$, we define $D^\alpha := \frac{\partial^{\alpha_1}}{\partial x_1^{\alpha_1}} \cdots \frac{\partial^{\alpha_n}}{\partial x_n^{\alpha_n}}$ and $|\alpha| := \sum_{i=1}^n \alpha_i$. By u' or $\partial_t u$ we denote the time derivative of a function

1

u, while ∇u denotes the gradient of the function u and $\nabla^2 u$ the matrix of second order derivatives.

For an operator A, $D(A)$ denotes the domain. We sometimes write $(A, D(A))$ for an operator, but if the domain is clear from the context, we just write A. rg A and ker A stand for the range and the kernel of the operator A, the adjoint operator is denoted by A'. $\sigma(A)$ and $\rho(A)$ denote the spectrum and the resolvent set of A, respectively. For $\lambda \in \rho(A)$, $R(\lambda, A)$ denotes the resolvent of the operator A at λ. The Laplacian Δ, is the sum of the pure second derivatives, i.e. $\Delta u := \sum_{i=1}^{n} D_{ii} u$.

Throughout this thesis a domain Ω is an open subset of $\mathbb{R}^n$ for some $n \in \mathbb{N}$ and $\partial \Omega$ denotes its boundary. For the Lebesgue spaces over a domain Ω we will use the standard notation $L^p(\Omega)$, $1 \leq p \leq \infty$, and the corresponding norms will be denoted by $\|\cdot\|_{L^p(\Omega)}$, or, when the domain is clear, simply by $\|\cdot\|_p$. $L^p_{\mathrm{loc}}(\Omega)$ is the space of functions on Ω which are locally, i.e. on each compact set in Ω, integrable in p-th power. For an exponent $p \in (1, \infty)$ we denote the conjugate exponent by p', i.e. $\frac{1}{p} + \frac{1}{p'} = 1$.

Furthermore, for $m \in \mathbb{N}_0$, $C^m(\Omega)$ denotes the space of m-times continuously differentiable functions while for $0 < \alpha \leq 1$,

$$C^{m,\alpha}(\Omega) := \left\{ f \in C^m(\Omega) : \right.$$

$$\left. \sup \left\{ \frac{|D^s f(x) - D^s f(y)|}{|x-y|^\alpha} : |s| = m, \ x, y \in \Omega, \ x \neq y \right\} < \infty \right\}.$$

$C_c^\infty(\Omega) = \mathcal{D}(\Omega)$ is the space of smooth functions with compact support in the set Ω and $\mathcal{D}'(\Omega)$ denotes the space of continuous functionals on $\mathcal{D}(\Omega)$. $\mathcal{S}(\mathbb{R}^n)$ denotes the space of Schwartz functions on $\mathbb{R}^n$ and $\mathcal{S}'(\mathbb{R}^n)$ its dual space.

By $\mathcal{F}f$ we denote the Fourier transform of a function $f \in \mathcal{S}'(\mathbb{R}^n)$. For $f \in \mathcal{S}(\mathbb{R}^n)$, it is defined by

$$\mathcal{F}f(\xi) := \frac{1}{(2\pi)^{n/2}} \int_{\mathbb{R}^n} e^{-ix\xi} f(x) \, \mathrm{d}x,$$

and for $f \in \mathcal{S}'(\mathbb{R}^n)$, we define $\langle \varphi, \mathcal{F}f \rangle := \langle \mathcal{F}\varphi, f \rangle$ for all $\varphi \in \mathcal{S}(\mathbb{R}^n)$. The notation $\hat{f}$ is reserved for the Laplace transform of functions f, defined by

$$\hat{f}(\lambda) := \int_0^\infty e^{-\lambda t} f(t) \, \mathrm{d}t.$$

At this point, we do not want to elaborate for which functions the Laplace transform exists. Suffice it to say, that the definition makes sense for any exponentially bounded function f whenever Re λ is sufficiently large. Both notations, i.e. $\mathcal{F}f$ and $\hat{f}$ are also used for the Fourier and Laplace transforms of Banach-space-valued functions f, respectively.

Given an angle $\theta \in (0, \pi]$, we define a sector in the complex plane by $\Sigma_\theta :=$ $\{\lambda \in \mathbb{C} : |\arg \lambda| < \theta\} \setminus \{0\}$, where $\arg \lambda$ denotes the argument of the complex number λ.

For any set M, $\overline{M}$ and M^c denote the closure of M and its complement, respectively, χ_M denotes the characteristic function of the set M.

Finally, the letter C stands for a constant which may change from line to line.

1.2 Function spaces on Lipschitz domains

In this section, we introduce the domains in $\mathbb{R}^n$ and the function spaces we will be working with and state some of their basic properties which will be used later on.

1.2.1 Lipschitz domains

Many known results for differential equations in domains rely on a localisation technique which reduces the problem to a problem in the half-space. This method requires the boundary to be sufficiently smooth (e.g. of class C^2, i.e. the boundary must be locally described by a C^2-function). The domains we will be working in, Lipschitz domains, do not have this property. Roughly speaking, a Lipschitz domain is a domain which can have corners and edges. Mathematically, it is a domain whose boundary can be locally described by a Lipschitz continuous function. This is made precise in the following definition.

Definition 1.2.1. *Let $\Omega \subseteq \mathbb{R}^n$ be a domain. It is called a* Lipschitz do-main *if there exists $M > 0$ so that every point on the boundary of Ω has a neighbourhood U such that, after an affine change of coordinates, $\partial\Omega \cap U$ is*

described by the equation $x_n = \varphi(x_1, ..., x_{n-1})$ where φ is a Lipschitz continuous function with Lipschitz constant bounded by M and $\Omega \cap U = \{x \in U : x_n > \varphi(x_1, ..., x_{n-1})\}$.

Remarks 1.2.2. 1. The requirement that the Lipschitz constants of the functions describing the boundary be uniformly bounded is not always made. Often, these domains are referred to as strongly Lipschitz. As we only consider domains of this type, we will only use the term Lipschitz domain. For bounded domains, or, more generally, Lipschitz domains with compact boundary, this distinction is unnecessary (cf. [Fra79, Theorem 3.5]).

2. Every Lipschitz domain satisfies both the cone and the segment property (cf. [Gri85, Section 1.2.2]).

3. A domain is called *of class* $C^{k,\alpha}$, $k \in \mathbb{N}_0$, $0 < \alpha \leq 1$, if all the functions φ in Definition 1.2.1 can be chosen to be $C^{k,\alpha}$-functions.

4. For Lipschitz continuous functions φ, we have $\nabla \varphi \in L^\infty_{\text{loc}}$, in particular, the gradient is defined almost everywhere (cf. [Alt85, Satz 8.5.2]).

1.2.2   Non-homogeneous Sobolev spaces

We will now define the function spaces in which we will be working for most of the time. These are the Sobolev spaces.

Definition 1.2.3. *Let Ω be a domain. Let $1 \leq p \leq \infty$ and $k \in \mathbb{N}$, then the Sobolev space of order k over $L^p(\Omega)$ is defined by*

$$W^{k,p}(\Omega) := \{f \in L^p(\Omega) : D^\alpha f \in L^p(\Omega) \text{ for } |\alpha| \leq k\},$$

where $D^\alpha f$ is to be understood as the derivative in the sense of distributions. On $W^{k,p}(\Omega)$ we introduce the norm

$$\|f\|_{W^{k,p}(\Omega)} := \left(\sum_{|\alpha| \leq k} \|D^\alpha f\|^p_{L^p(\Omega)} \right)^{\frac{1}{p}}.$$

For $s \in \mathbb{R}_+$, we define the fractional order Sobolev spaces via the complex interpolation method (cf. [BL76]).

$$W^{s,p}(\Omega) := \left[W^{k,p}(\Omega), W^{k+1,p}(\Omega) \right]_\theta,$$

where $k \in \mathbb{N}_0$ and $s = k + \theta$ for $\theta \in [0,1)$.

The closure of the space of $C_c^\infty(\Omega)$-functions in $W^{s,p}(\Omega)$ will be denoted by $W_0^{s,p}(\Omega)$.

We define the Sobolev spaces of negative order *as the dual spaces to $W_0^{s,p}(\Omega)$, i.e.*

$$W^{-s,p}(\Omega) := \left(W_0^{s,p'}(\Omega) \right)'.$$

Remarks 1.2.4. Often the fractional order Sobolev spaces are defined differently using Hölder continuity (see [Ada75] or [Gri85]). This leads to a different scale of Sobolev spaces. Using complex interpolation has the advantage that the fractional order Sobolev spaces coincide with the Bessel potential spaces on Lipschitz domains.

Definition 1.2.5. *Let $1 \leq p \leq \infty$ and $\alpha \in \mathbb{R}$, then the* Bessel potential spaces *are defined by*

$$L_\alpha^p(\mathbb{R}^n) := \left\{ f \in \mathcal{S}'(\mathbb{R}^n) : (I - \Delta)^{\alpha/2} f \in L^p(\mathbb{R}^n) \right\}$$

with norm

$$\|f\|_{L_\alpha^p(\mathbb{R}^n)} := \left\| (I - \Delta)^{\alpha/2} f \right\|_{L^p(\mathbb{R}^n)}.$$

Here, $(I - \Delta)^{\alpha/2} f$ is defined via the Fourier transform as

$$(I - \Delta)^{\alpha/2} f := \mathcal{F}^{-1} \left((1 + |\cdot|^2)^{\alpha/2} \mathcal{F} f \right).$$

On domains $\Omega \subset \mathbb{R}^n$, we define the Bessel potential spaces via restriction. Let $R_\Omega f$ denote the restriction of f to Ω. For $\alpha \geq 0$ define

$$L_\alpha^p(\Omega) := R_\Omega L_\alpha^p(\mathbb{R}^n)$$

with the quotient norm

$$\|f\|_{L_\alpha^p(\Omega)} := \inf\{\|g\|_{L_\alpha^p(\mathbb{R}^n)} : R_\Omega g = f\}.$$

For $1 < p < \infty$ and $\alpha \in \mathbb{R}$, define

$$L_{\alpha,0}^p(\Omega) := \{f \in L_\alpha^p(\mathbb{R}^n), \ \mathrm{supp} f \subset \overline{\Omega}\}$$

with norm

$$\|f\|_{L_{\alpha,0}^p(\Omega)} := \|f\|_{L_\alpha^p(\mathbb{R}^n)}.$$

Finally, for $1 < p < \infty$ and $\alpha \in \mathbb{R}_+$, define $L_{-\alpha}^p(\Omega)$ as the space of linear functionals on $C_c^\infty(\Omega)$ such that the norm

$$\|g\|_{L_{-\alpha}^p(\Omega)} := \sup\{|g(f)| : f \in C_c^\infty(\Omega), \|f\|_{L_\alpha^{p'}(\Omega)} \le 1\}$$

is finite.

Remarks 1.2.6. 1. From the theory of Fourier multipliers (cf. Section 1.3.4), it follows that, for $k \in \mathbb{N}$ and $1 \le p < \infty$, we have $W^{k,p}(\mathbb{R}^n) = L_k^p(\mathbb{R}^n)$.

2. If Ω is a Lipschitz domain, we can extend functions in $W^{k,p}(\Omega)$ to $W^{k,p}(\mathbb{R}^n)$ using an extension operator (cf. [Cal61] or Theorem 1.2.9). In this way, we obtain $W^{k,p}(\Omega) = L_k^p(\Omega)$ for Lipschitz domains.

3. As the Bessel potential spaces form a complex interpolation scale for $\alpha \ge 0$ (cf. [Tri78, 4.3.1]), they coincide with the spaces gained by the complex interpolation method from the spaces $W^{k,p}(\Omega)$, $k \in \mathbb{N}$, for Lipschitz domains Ω.

4. For Lipschitz domains, we have that $C_c^\infty(\Omega)$ is dense in $L_{\alpha,0}^p(\Omega)$ for any $1 < p < \infty$ and $-\infty < \alpha < \infty$ (cf. [JK95, Remark 2.7]). Therefore, $L_{s,0}^p(\Omega) = W_0^{s,p}(\Omega)$ for $s \ge 0$.

5. It may seem unnecessary to introduce both the Sobolev spaces and the Bessel potential spaces considering that they are identical in all cases we treat. However, the norms may differ and although the norms are of course equivalent for a fixed domain Ω, the constant $C = C(\Omega)$ in the estimate

$$\frac{1}{C} \|u\|_{L_\alpha^p(\Omega)} \le \|u\|_{W^{\alpha,p}(\Omega)} \le C \|u\|_{L_\alpha^p(\Omega)}$$

may depend on the domain. This will be investigated in more detail when it becomes important later on.

1.2.3   Density, extension and embedding theorems

It is often very useful to approximate functions in Sobolev spaces by smooth functions. The following standard result can be found e.g. in the books [Agm65, Theorem 2.1] or [Gri85, Theorem 1.4.2.1].

Theorem 1.2.7. *Let Ω be a Lipschitz domain in $\mathbb{R}^n$. Then $C_c^\infty(\overline{\Omega}) := \{R_\Omega u : u \in C_c^\infty(\mathbb{R}^n)\}$ is dense in $W^{k,p}(\Omega)$ for all $k \in \mathbb{N}$ and $1 < p < \infty$.*

Remarks 1.2.8. 1. The requirement on the boundary in Theorem 1.2.7 can be relaxed to Ω having a continuous boundary ([Gri85, Theorem 1.4.2.1]) or that Ω have the segment property ([Agm65, Theorem 2.1]). That these two conditions are in fact equivalent is shown in [Fra79].

2. The same result result can be shown for Lipschitz domains for non-integer values of k. This is done in [JK95, Proposition 2.9].

If Ω is a domain in $\mathbb{R}^n$ such that a continuous extension operator E : $W^{s,p}(\Omega) \to W^{s,p}(\mathbb{R}^n)$ exists, then many results that hold in $W^{s,p}(\mathbb{R}^n)$ can be transferred to $W^{s,p}(\Omega)$. On Lipschitz domains we have the following result:

Theorem 1.2.9. *Let Ω be a Lipschitz domain in $\mathbb{R}^n$. Then there exists an operator E mapping functions on Ω to functions on $\mathbb{R}^n$ such that*

a) $E(f)|_\Omega = f$,

b) $E : W^{k,p}(\Omega) \to W^{k,p}(\mathbb{R}^n)$ is continuous for all $k \in \mathbb{N}_0$ and $1 \leq p \leq \infty$.

Remarks 1.2.10. 1. Because of the property stated in a), E is called an extension operator.

2. Stein ([Ste70, Theorem VI.3.5]) proves this result for the class of domains with a so-called minimally smooth boundary which includes Lipschitz domains. He also gives an example proving that singularities given by boundary functions $\varphi(x') = |x'|^\gamma$, $\gamma < 1$, do not allow for continuous extension operators, and so the class of Lipschitz domains is 'in the nature of the best possible' where results as in Theorem 1.2.9 can be expected.

We now want to present the main results on embeddings of Sobolev spaces. The classical Sobolev embedding theorem (cf. [Tri78, Theorem 2.8.1][1]) states that

$$W^{s,p}(\mathbb{R}^n) \hookrightarrow W^{t,q}(\mathbb{R}^n)$$

---

[1]Note that Triebel states the result for the spaces $F^s_{p,r}(\mathbb{R}^n)$. For $r = 2$, these spaces coincide with the Bessel potential spaces (cf. [Tri78, Theorem 2.3.3]) and therefore with the Sobolev spaces we introduced. Other results as in [Ada75, Theorem 5.4] refer to a different scale of Sobolev spaces, namely those gained by the real interpolation method.

for $t \leq s$, $q \geq p$ with $s - \frac{n}{p} = t - \frac{n}{q}$ and

$$W^{s,p}(\mathbb{R}^n) \hookrightarrow C^{k,\alpha}(\mathbb{R}^n)$$

for $k < s - \frac{n}{p} < k + 1$, and $\alpha = s - k - \frac{n}{p}$, $k \in \mathbb{N}_0$.

Using the extension operator from Theorem 1.2.9, these results can also be shown for domains.

Theorem 1.2.11. *Let Ω be a Lipschitz domain in $\mathbb{R}^n$. Then*

$$W^{s,p}(\Omega) \hookrightarrow W^{t,q}(\Omega)$$

for $t \leq s$, $q \geq p$ with $s - \frac{n}{p} = t - \frac{n}{q}$ and

$$W^{s,p}(\Omega) \hookrightarrow C^{k,\alpha}(\Omega)$$

for $k < s - \frac{n}{p} < k + 1$, and $\alpha = s - k - \frac{n}{p}$, $k \in \mathbb{N}_0$.

If the domain Ω is bounded, we get compact embeddings, in particular we have (cf. [Ada75, Theorem 6.2] or [Alt85, Theorem A 6.4])

Theorem 1.2.12 (Rellich). *Let Ω be a bounded Lipschitz domain in $\mathbb{R}^n$, $1 \leq p < \infty$ and $k \in \mathbb{N}$. Then the embedding*

$$W^{k,p}(\Omega) \hookrightarrow W^{k-1,p}(\Omega)$$

is compact.

This allows us to estimate intermediate derivatives in the following way:

Lemma 1.2.13 (Ehrling). *Let Ω be a bounded Lipschitz domain in $\mathbb{R}^n$ and $1 \leq p < \infty$. Then for all $\varepsilon > 0$ there exists a constant $C_\varepsilon > 0$ such that*

$$\left\| \nabla u \right\|_p \leq \varepsilon \left\| \nabla^2 u \right\|_p + C_\varepsilon \left\| u \right\|_p$$

for all $u \in W^{2,p}(\Omega)$.

Proof. See [Ada75, Theorem 4.14]. $\square$

A similar estimate for intermediate derivatives is also given by the Gagliardo-Nirenberg inequality (cf. [Maz85, Section 1.4.8]).

Lemma 1.2.14 (Gagliardo-Nirenberg). *Let Ω be a bounded Lipschitz domain in $\mathbb{R}^n$. Let $j \leq l$. Then there exists a constant C such that for all $u \in W^{l,p}(\Omega)$ we have*

$$(1.1) \qquad \left\| \nabla^j u \right\|_q \leq C (\left\| \nabla^l u \right\|_p + \left\| u \right\|_r)^a \left\| u \right\|_r^{1-a},$$

where $p \geq 1$, $1/q = j/n + a(1/p - l/n) + (1-a)/r$ for all $a \in [j/l, 1]$ unless $1 < p < \infty$ and $l - j - n/p \in \mathbb{N}_0$ when (1.1) holds for $a \in [j/l, 1)$.

Another useful estimate is by given Poincaré's inequality. This allows us to estimate norms of certain functions by the norm of their derivative. A proof can be found in [Nec67, Section 2.7].

Theorem 1.2.15 (Poincaré's inequality). *Let Ω be a bounded Lipschitz domain in $\mathbb{R}^n$ and $1 \leq p < \infty$. Then there exists a constant C depending only on the diameter of Ω such that*

$$\left\| u \right\|_p \leq C \left\| \nabla u \right\|_p$$

for all $u \in W_0^{1,p}(\Omega)$.

Remarks 1.2.16. Theorem 1.2.15 holds when Ω is bounded in one direction only.

An alternative formulation is given by

Theorem 1.2.17 (Generalised Poincaré inequality). *Suppose Ω is a bounded Lipschitz domain in $\mathbb{R}^n$, $k \in \mathbb{N}$ and that $1 \leq p < \infty$. Then there exists a constant C such that*

$$\left\| u \right\|_{W^{k,p}} \leq C \left(\left\| \nabla^k u \right\|_p^p + \sum_{|\alpha| \leq k-1} \left| \int_\Omega D^\alpha u \right|^p \right)^{\frac{1}{p}}$$

for all $u \in W^{k,p}(\Omega)$.

Proof. See [Kuf77, Theorem 5.11.3] or [Nec67, Section 2.7]. $\qquad\qquad\square$

1.2.4   Boundary values of functions in Sobolev spaces

Let Ω be a bounded Lipschitz domain in $\mathbb{R}^n$ and $1 \leq p \leq \infty$. Then there exists a unique operator

$$\mathrm{tr} : W^{1,p}(\Omega) \to L^p(\partial\Omega)$$

such that $\mathrm{tr}\, u = u|_{\partial\Omega}$ for all $u \in W^{1,p}(\Omega) \cap C(\overline{\Omega})$. We call tr the trace operator. This is a standard construction which can be found for example in [Alt85, A 6.6] or [Nec67, Section 2.4] and which also works for domains with a compact boundary.

To define a trace on unbounded Lipschitz domains, we use the following approximation procedure. Let $R > 0$ and C_R be a cylinder of height R with axis in the x_n-direction and with a base of diameter R. Set $\Omega_R := \Omega \cap C_R$ and $u_R = u|_{\Omega_R}$. If $u \in W^{1,p}_{\mathrm{loc}}(\Omega)$, then $u_R \in W^{1,p}(\Omega_R)$ and the trace operator tr_R on Ω_R is defined. For $R_1 < R_2$ we get that

$$\mathrm{tr}_{R_1}(u_{R_1}) = \mathrm{tr}_{R_2}(u_{R_2}) \text{ on } \partial\Omega \cap C_{R_1},$$

by continuity of the trace operator as this holds for $u \in C(\overline{\Omega_{R_2}})$. Therefore, we can construct a function $\mathrm{tr}\, u \in L^p_{\mathrm{loc}}(\partial\Omega)$ such that

$$\mathrm{tr}\, u|_{\partial\Omega \cap C_R} = \mathrm{tr}_R\, u_R|_{\partial\Omega \cap C_R} \text{ for all } R > 0.$$

This operator defined for $u \in W^{1,p}_{\mathrm{loc}}(\Omega)$ will be called the trace of u on $\partial\Omega$. For $u \in C(\overline{\Omega})$, we have $\mathrm{tr}\, u = u|_{\partial\Omega}$ as this is the case for all operators tr_R.

Similarly, it is possible to define a continuous trace operator $\mathrm{tr} : W^{\alpha,p}(\Omega) \to L^p(\partial\Omega)$ for $\alpha > 1/p$. The spaces $W^{\alpha,p}_0(\Omega)$ can then be thought of as the spaces of those functions in $W^{\alpha,p}(\Omega)$ which vanish at the boundary in the sense that their trace is zero.

Proposition 1.2.18. *Let Ω be a bounded Lipschitz domain and $1/p < \alpha < 1 + 1/p$. Then*

$$W^{\alpha,p}_0(\Omega) = \{f \in W^{\alpha,p}(\Omega) : \mathrm{tr} f = 0 \text{ on } \partial\Omega\}.$$

Proof.   Cf. [JK95, Proposition 3.3].   □

Remarks 1.2.19. Some remarks on the range of allowed α seem to be in order.

1. For $\alpha \leq 1/p$, the trace operator cannot be defined as a continuous operator from $W^{\alpha,p}(\Omega)$ to $L^p(\partial\Omega)$. In fact, in bounded Lipschitz domains, for $0 \leq \alpha \leq 1/p$, $C_c^\infty(\Omega)$ is dense in $W^{\alpha,p}(\Omega)$, so $W_0^{\alpha,p}(\Omega) = W^{\alpha,p}(\Omega)$ (cf. [JK95, Section 3]).

2. For $\alpha > 1 + 1/p$, it is possible to define the trace of the first derivatives of functions in $W^{\alpha,p}(\Omega)$. Therefore, at least philosophically, we must expect not just the trace of the function, but also the trace of its gradient, to vanish for functions in $W_0^{\alpha,p}(\Omega)$ when $\alpha > 1 + 1/p$.

Next, we define the outer unit normal for Lipschitz domains.

Definition 1.2.20. *Let Ω be a Lipschitz domain and $x \in \partial\Omega$. Let φ be the Lipschitz function describing $\partial\Omega$ in a neighbourhood of x. Then $\nabla\varphi$ is defined almost everywhere (see Remark 1.2.2). Let $x = (x', \varphi(x'))$ and e_i be the i-th unit vector. Then*

$$N(x) = \left(1 + |\nabla\varphi(x')|^2\right)^{-\frac{1}{2}} \left(\sum_{i=1}^{n-1} D_i\varphi(x')e_i - e_n\right)$$

is the outer unit normal *in x.*

Remarks 1.2.21. This definition is independent of the local representation of $\partial\Omega$.

It is very reassuring to know that integration by parts works as expected in Lipschitz domains. We therefore state these results which can be found e.g. in [Gri85, Theorem 1.5.3.1].

Proposition 1.2.22. *For bounded Lipschitz domains $\Omega \subseteq \mathbb{R}^n$ we have*

$$(1.2) \qquad \int_\Omega D_i u \; v = -\int_\Omega u \; D_i v + \int_{\partial\Omega} uvN_i$$

for $u \in W^{1,p}(\Omega)$, $v \in W^{1,p'}(\Omega)$ and $1 \leq i \leq n$.

Remarks 1.2.23. The boundary integral on the right hand side of (1.2) is to be understood in the sense of trace. Note that, as Ω has a compact boundary, $\operatorname{tr}: W^{1,q}(\Omega) \to L^q(\partial\Omega)$ is continuous for all $1 < q < \infty$ and so the boundary integral exists.

For unbounded domains this gives the following result:

Corollary 1.2.24. *Let $\Omega \subseteq \mathbb{R}^n$ be any Lipschitz domain. Then for $u \in W_0^{1,p}(\Omega)$ and $v \in W^{1,p'}(\Omega)$ we have*

$$\int_\Omega D_i u \, v = -\int_\Omega u \, D_i v.$$

Proof. It is sufficient to show the statement for $u \in C_c^\infty(\Omega)$. Let $\Omega_R = \Omega \cap B_R$. Then Proposition 1.2.22 implies

$$\int_{\Omega_R} D_i u \, v = -\int_{\Omega_R} u \, D_i v + \int_{\partial\Omega_R} uvN_i.$$

Choosing $R > 0$ sufficiently large that supp $u \subseteq \Omega_R$, the last term is zero and we get

$$\int_\Omega D_i u \, v = \lim_{R\to\infty} \int_{\Omega_R} D_i u \, v = -\lim_{R\to\infty}\int_{\Omega_R} u \, D_i v = -\int_\Omega u \, D_i v. \qquad \square$$

Neumann boundary conditions for the Laplacian are to be understood in the following way:

Definition 1.2.25. *For a domain Ω, $1 < p < \infty$ and $u \in W^{1,p}(\Omega)$ with $\Delta u \in L^p(\Omega)$, we say that $\frac{\partial u}{\partial N} = 0$ iff*

$$(1.3) \qquad \int_\Omega \Delta u \varphi = -\int_\Omega \nabla u \nabla \varphi$$

for any $\varphi \in W^{1,p'}(\Omega)$.

Remarks 1.2.26. *1. $\Delta u \in L^p(\Omega)$ is to be understood in the sense of distributions.*

2. For bounded Lipschitz domains, $u \in W^{2,p}(\Omega)$ and $\varphi \in W^{1,p'}(\Omega)$, Proposition 1.2.22 yields

$$\int_\Omega \Delta u \, \varphi = -\int_\Omega \nabla u \nabla \varphi + \sum_{i=1}^n \int_{\partial\Omega} D_i u \, \varphi \, N_i$$

and the last term is zero for any $\varphi \in W^{1,p'}(\Omega)$ iff

$$\sum_{i=1}^n D_i u \, N_i = \frac{\partial u}{\partial N} = 0.$$

More generally, for $\varphi \in W^{1,p'}(\Omega)$ we have tr $\varphi \in B_{p',p'}^{1-1/p'}(\partial\Omega)$.[2] In fact, tr : $W^{1,p'}(\Omega) \to B_{p',p'}^{1-1/p'}(\partial\Omega)$ is surjective (cf. [JK95, Theorem 3.1]). So for $u \in W^{1,p}(\Omega)$ with $\Delta u \in L^p(\Omega)$ the condition (1.3) corresponds to tr $\frac{\partial u}{\partial N} \in (B_{p',p'}^{1-1/p'}(\partial\Omega))' = B_{p,p}^{-1/p}(\partial\Omega)$ and tr $\frac{\partial u}{\partial N} = 0$.

1.2.5   Homogeneous Sobolev spaces

We now introduce the homogeneous Sobolev spaces. These will play an important role in unbounded domains. In the following, we let $\Omega \subseteq \mathbb{R}^n$ be a Lipschitz domain.[3]

Definition 1.2.27. Let $k \in \mathbb{N}$ and

$$\widehat{W}^{k,p}(\Omega) := \{ f \in L^p_{\mathrm{loc}}(\Omega) : D^\alpha f \in L^p(\Omega) \text{ for } |\alpha| = k \}.$$

On $\widehat{W}^{k,p}(\Omega)$ we introduce the semi-norm

$$\|u\|_{k,p}^{\wedge} = \left(\sum_{|\alpha|=k} \|D^\alpha u\|_p^p \right)^{\frac{1}{p}}.$$

Then $\widehat{W}^{k,p}(\Omega)$ is called the homogeneous Sobolev space of order k.

Furthermore, we introduce the space of functions vanishing at the boundary

$$\widehat{W}_0^{1,p}(\Omega) := \{ f \in L^p_{\mathrm{loc}}(\Omega) : \nabla f \in L^p(\Omega) \text{ and } \mathrm{tr}\, u = 0 \}$$

and the space

$$\widehat{\mathcal{W}}^{k,p}(\Omega) := \widehat{W}^{k,p}(\Omega)_{/P_{k-1}}$$

where P_{k-1} denotes the space of polynomials of degree at most $k-1$.

As $P_{k-1} \subseteq \widehat{W}^{k,p}(\Omega)$, and $\|f\|_{k,p}^{\wedge} = 0$ for all $f \in P_{k-1}$, the space $\widehat{W}^{k,p}(\Omega)$ cannot be a Banach space. By factoring out the polynomials however, we obtain a complete normed space.

Proposition 1.2.28. $\widehat{\mathcal{W}}^{k,p}(\Omega)$ is a reflexive Banach space for all $k \in \mathbb{N}$.

---

[2]Here, $B_{p,q}^s(\Omega)$ denotes the Besov space (see [BL76] or [Tri78]).

[3]The definition obviously makes sense for more general domains, however for some of the proofs given later in this section, we use that Ω is a Lipschitz domain.

Proof. See [Maz85, 1.1.13] or [Nec67, Section 2.7] for the proof of completeness, reflexivity follows from the fact that $\widehat{W}^{k,p}(\Omega)$ is isometric to a closed
subspace of $L^p(\Omega)^n$. □

As we will be considering boundary value problems on unbounded domains
and the spaces $\widehat{W}^{k,p}(\Omega)$ aren't Banach spaces, we need to introduce a trace
operator on the Banach space $\widehat{\mathcal{W}}^{1,p}(\Omega)$.

Let $u \in \widehat{W}^{1,p}(\Omega)$, denote its equivalence class in $\widehat{\mathcal{W}}^{1,p}(\Omega)$ by $\tilde{u}$ and let $\widetilde{\mathrm{tr}\, u}$
denote the equivalence class of $\mathrm{tr}\, u$ in $L^p_{\mathrm{loc}}(\partial\Omega)_{/\mathrm{const}}$. Then define $\widetilde{\mathrm{tr}}$ on
$\widehat{\mathcal{W}}^{1,p}(\Omega)$ by

$$\widetilde{\mathrm{tr}}\, \tilde{u} := \widetilde{\mathrm{tr}\, u}.$$

We can now introduce the following space of functions vanishing at the
boundary modulo constants:

$$\widehat{\mathcal{W}}^{1,p}_0(\Omega) \ := \ \{\tilde{u} \in \widehat{\mathcal{W}}^{1,p}(\Omega) : \widetilde{\mathrm{tr}}\, \tilde{u} = 0\}.$$

We also introduce the dual space

$$\widehat{\mathcal{W}}^{-1,p}(\Omega) \ := \ (\widehat{\mathcal{W}}^{1,p'}_0(\Omega))'.$$

Proposition 1.2.29. *The space $\widehat{\mathcal{W}}^{1,p}_0(\Omega)$ is a reflexive Banach space.*

Proof. We need to show that $\widehat{\mathcal{W}}^{1,p}_0(\Omega)$ is a closed subspace of $\widehat{\mathcal{W}}^{1,p}(\Omega)$. To
see this, we let $\tilde{\phi} \in \widetilde{\mathrm{tr}}(\widehat{\mathcal{W}}^{1,p}(\Omega))$ and set

$$\left\|\tilde{\phi}\right\|_\wedge := \inf\{\left\|\nabla u\right\|_p : u \in \widehat{W}^{1,p}(\Omega) \text{ and } \tilde{\phi} = \widetilde{\mathrm{tr}}\, \tilde{u}\}.$$

We now have to show that $\left\|\cdot\right\|_\wedge$ is a norm on $\widetilde{\mathrm{tr}}(\widehat{\mathcal{W}}^{1,p}(\Omega))$. Then, as by
definition of $\left\|\cdot\right\|_\wedge$, the operator $\widetilde{\mathrm{tr}} : \widehat{\mathcal{W}}^{1,p}(\Omega) \to \widetilde{\mathrm{tr}}(\widehat{\mathcal{W}}^{1,p}(\Omega))$ is continuous,
$\mathrm{Ker}\, \widetilde{\mathrm{tr}} = \widehat{\mathcal{W}}^{1,p}_0(\Omega)$ is a closed subspace of $\widehat{\mathcal{W}}^{1,p}(\Omega)$.

Obviously, for $\tilde{\phi} \in \widetilde{\mathrm{tr}}(\widehat{\mathcal{W}}^{1,p}(\Omega))$ and $\lambda \in \mathbb{C}$ we have that

$$\left\|\lambda\tilde{\phi}\right\|_\wedge = |\lambda| \left\|\tilde{\phi}\right\|_\wedge$$

and if $\tilde{\phi} = \widetilde{\mathrm{tr}}\, \tilde{u}$ and $\tilde{\psi} = \widetilde{\mathrm{tr}}\, \tilde{v}$, then $\tilde{\phi} + \tilde{\psi} = \widetilde{\mathrm{tr}}\, (\tilde{u} + \tilde{v})$, so

$$\left\|\tilde{\phi} + \tilde{\psi}\right\|_\wedge \leq \left\|\tilde{\phi}\right\|_\wedge + \left\|\tilde{\psi}\right\|_\wedge .$$

It remains to show that

$$\left\|\tilde{\phi}\right\|_{\wedge} = 0 \;\Rightarrow\; \tilde{\phi} = 0.$$

Suppose we have a sequence $(u_n) \subseteq \widehat{W}^{1,p}(\Omega)$ such that for all $n \in \mathbb{N}$, we have $\|\nabla u_n\|_p \leq 1/n$ and $\tilde{\phi} = \widetilde{\mathrm{tr}}(\tilde{u}_n)$. Let $R > 0$, $\Omega_R = \Omega \cap B_R$ and let

$$c_n = \frac{1}{|\Omega_R|} \int_{\Omega_R} u_n.$$

The generalised Poincaré inequality (Theorem 1.2.17) then gives the following estimate

$$\|u_n - c_n\|_{W^{1,p}(\Omega_R)} \leq \frac{C}{n},$$

and $\mathrm{tr}(u_n - c_n) \to 0$ in $L^p(\partial\Omega_R)$ implies $\widetilde{\mathrm{tr}}\, \tilde{c}_n \to \tilde{\phi}$, but $\widetilde{\mathrm{tr}}\, \tilde{c}_n = 0$, so $\tilde{\phi} = 0$. $\qquad\square$

The data f will sometimes be chosen in the space $\widehat{\mathcal{W}}^{-1,p}(\Omega)$. It is therefore useful to know that smooth functions with compact support are dense in this space.

Proposition 1.2.30. $C_c^\infty(\Omega) \cap \widehat{\mathcal{W}}^{-1,p}(\Omega)^4$ *is dense in* $\widehat{\mathcal{W}}^{-1,p}(\Omega)$.

Proof. Note that

$$\|f\|_{-1,p}^{\wedge} = \sup_{0 \neq h \in \widehat{\mathcal{W}}_0^{1,p'}(\Omega)} \frac{|f(h)|}{\|\nabla h\|_{p'}}.$$

If $C_c^\infty(\Omega) \cap \widehat{\mathcal{W}}^{-1,p}(\Omega)$ were not dense in $\widehat{\mathcal{W}}^{-1,p}(\Omega)$, the Hahn-Banach Theorem and Proposition 1.2.29 would imply the existence of $0 \neq h \in \widehat{\mathcal{W}}_0^{1,p'}(\Omega)$ such that $\int fh = 0$ for all $f \in C_c^\infty(\Omega) \cap \widehat{\mathcal{W}}^{-1,p}(\Omega)$. Let $g \in (C_c^\infty(\Omega))^n$ and set $f := -\mathrm{div}\, g$. As

$$\left|\int fv\right| = \left|\int \mathrm{div}\, g\, v\right| = \left|\int g\nabla v\right| \leq \|g\|_p \|v\|_{1,p'}^{\wedge}$$

for all $v \in \widehat{\mathcal{W}}_0^{1,p'}(\Omega)$, we have that $f \in C_c^\infty(\Omega) \cap \widehat{\mathcal{W}}^{-1,p}(\Omega)$. Then

$$\int g\nabla h = \int fh = 0$$

implies that $\nabla h = 0$ in $L^{p'}(\Omega)$ which gives a contradiction. $\qquad\square$

---

[4]Note that $C_c^\infty(\Omega)$ is not contained in $\widehat{\mathcal{W}}^{-1,p}(\Omega)$ as if f is a function that lies in $\widehat{\mathcal{W}}^{-1,p}(\Omega)$, it must have mean 0.

1.3   Operator theory

In this section we briefly review some of the main concepts from operator theory that will be used later in this thesis. For a more detailed treatment of these topics we refer to [EN00] or [ABHN01] for C_0-semigroups and [DHP03] for H^∞-calculus and maximal regularity.

1.3.1 C_0-semigroups

C_0-semigroups play an important role in the theory of parabolic equations. As we will see below, in many cases, they allow us to find a classical solution to the abstract Cauchy problem

$$(1.4) \qquad\qquad \begin{cases} u'(t) = Au(t) & \text{for } t > 0, \\ u(0) = u_0 \end{cases}$$

in a Banach space X by studying spectral properties of the operator A. By a classical solution to (1.4), we mean a function $u \in C^1(\mathbb{R}_+, X)$ such that $u(t) \in D(A)$ for $t > 0$ and (1.4) holds. We now make some basic definitions.

Definition 1.3.1. *Let X, Y be Banach spaces. A function $T : \mathbb{R}_+ \to \mathcal{L}(X, Y)$ is called* strongly continuous *if the map $t \mapsto T(t)x$ from $\mathbb{R}_+$ into Y is continuous for all $x \in X$.*

A strongly continuous function $T : \mathbb{R}_+ \to \mathcal{L}(X)$ is called a C_0-semigroup if the following conditions are satisfied:

a) $T(0) = I$.

b) $T(t + s) = T(t)T(s)$ for all $t, s \geq 0$.

We now collect some basic results on C_0-semigroups. For proofs, we refer to [ABHN01, Section 3.1].

If T is a C_0-semigroup, then there exist constants $M \geq 1$ and $\omega \in \mathbb{R}$ such that

$$\|T(t)\| \leq Me^{\omega t}, \quad t \geq 0.$$

The infimum over all possible ω is called the *growth bound* $\omega(T)$ of the semigroup. We say that T is

a) *of negative type* if $\omega(T) < 0$,

b) *quasi-contractive* if $M = 1$ is possible and

c) *contractive* if $\|T(t)\| \leq 1$ for all $t \geq 0$.

Furthermore, for C_0-semigroups T, there exists an operator A and some $\lambda_0 \in \mathbb{R}$ such that $(\lambda_0, \infty) \subseteq \rho(A)$ and

$$\hat{T}(\lambda) = R(\lambda, A) \quad \text{for} \quad \lambda > \lambda_0.$$

We call the operator A the *generator* of the semigroup T.

If A is the generator of a C_0-semigroup and $u_0 \in D(A)$ then the unique classical solution to (1.4) is given by $u(t) = T(t)u_0$. Therefore, it is of great interest to determine those operators A that generate C_0-semigroups. One of the most celebrated theorems in this area is the generation theorem by Hille and Yoshida. It characterises generators of C_0-semigroups via resolvent estimates. We state the theorem for the contractive case.

Theorem 1.3.2 (Hille-Yosida Theorem, Contractive case). *For a linear operator A on a Banach space X, the following are equivalent:*

1. *A generates a contractive C_0-semigroup.*

2. *A is densely defined and for all $\lambda > 0$, we have $\lambda \in \rho(A)$ and*

$$\|\lambda R(\lambda, A)\| \leq 1.$$

Proof. See [EN00, Theorem II.3.5]. □

A different way of determining whether an operator A is the generator of a C_0-semigroup is given by the Lumer-Phillips Theorem. Before stating the theorem, we define the notion of dissipativity for linear operators.

Definition 1.3.3. *A linear operator A on X is called* dissipative *if*

$$\|(\lambda - A)x\| \geq \lambda \|x\|$$

for all $\lambda > 0$ and $x \in D(A)$.

This property can also be characterised in another way. For this, we first introduce the subdifferential of the norm at an element $x \in X$:

$$dN(x) := \{x^* \in X^* : \|x^*\| \leq 1, \langle x, x^* \rangle = \|x\|\}.$$

By the Hahn-Banach Theorem, the set $dN(x)$ is non-empty.

Lemma 1.3.4. *An operator A is dissipative iff for every $x \in D(A)$, there exists $x^* \in dN(x)$ such that*

$$\mathrm{Re}\,\langle Ax, x^* \rangle \leq 0.$$

Proof.   See [ABHN01, Lemma 3.4.2]. □

We are now ready to state the generation theorem due to Lumer and Phillips. For a proof, we refer to [ABHN01, Theorem 3.4.5].

Theorem 1.3.5 (Lumer, Phillips). *For a densely defined, dissipative operator A, the following are equivalent:*

1. *The closure $\overline{A}$ of A generates a contractive semigroup.*

2. *$\mathrm{rg}(\lambda - A)$ is dense in X for one/all $\lambda > 0$.*

We now consider families of operators that depend on a complex parameter.

Definition 1.3.6. *Let $\theta \in \left(0, \frac{\pi}{2}\right]$. A semigroup T is called* analytic of angle θ *if it has an analytic extension to the sector Σ_θ which is bounded on $\Sigma_{\theta'} \cap \{z \in \mathbb{C} : |z| \leq 1\}$ for all $0 < \theta' < \theta$.*
If T has a bounded analytic extension to $\Sigma_{\theta'}$ for each $\theta' \in (0, \theta)$, we call T a bounded analytic semigroup of angle θ.

Generators of analytic semigroups can also be characterised using resolvent estimates (see e.g. [ABHN01, Theorem 3.7.11]).

Theorem 1.3.7. *Let $\theta \in \left(0, \frac{\pi}{2}\right]$. For an operator A, the following are equivalent:*

1. *A generates a bounded analytic semigroup of angle θ.*

2. $\Sigma_{\theta+\frac{\pi}{2}} \subset \rho(A)$ and for all $0 < \theta' < \theta$ there exists a constant M such that

$$\|R(\lambda, A)\| \leq \frac{M}{|\lambda|}$$

for all $\lambda \in \Sigma_{\theta'+\frac{\pi}{2}}$.

Remarks 1.3.8. 1. An operator A generates an analytic semigroup iff there exists $c \geq 0$ such that $A - c$ generates a bounded analytic semigroup.

2. If A is the generator of a bounded C_0-semigroup T on the Banach space X, then T is a bounded analytic semigroup iff $T(t)x \in D(A)$ for all $x \in X$ and $t > 0$ and

$$\sup_{t>0} \|tAT(t)\|_X < \infty.$$

See [ABHN01, Theorem 3.7.19] for a proof.

3. If A is the generator of an analytic semigroup then $u(t) = T(t)u_0$ is the classical solution of (1.4) for any $u_0 \in X$ (see [ABHN01, Corollary 3.7.21]).

Another useful tool for determining whether an operator is the generator of a C_0-semigroup is given by perturbation theory. If A generates a C_0-semigroup and A is perturbed by an operator B which is in a sense small compared to A, then the perturbed operator $A + B$ can still generate a C_0-semigroup. We first make precise what we mean by an operator B being small compared to A.

Definition 1.3.9. *Let A be an operator on the Banach space X. An operator B is called* relatively A-bounded *if $\mathcal{D}(A) \subseteq \mathcal{D}(B)$ and there exist constants $a, b \in \mathbb{R}_+$ such that*

$$\|Bx\| \leq a \|Ax\| + b \|x\| \quad \text{for all } x \in \mathcal{D}(A).$$

The infimum of all possible constants a is called the A-bound or relative bound.

For perturbations by dissipative operators, we have the following result which is proved in [EN00, Theorem III.2.7]:

Theorem 1.3.10. *Let A be the generator of a contraction semigroup and assume B is dissipative and A-bounded with A-bound $a_0 < 1$. Then $(A + B, \mathcal{D}(A))$ generates a contraction semigroup.*

Remarks 1.3.11. In particular, the theorem shows that if the operator A is perturbed with a dissipative, A-bounded operator with A-bound $a_0 < 1$, then the growth bound of the semigroup remains unchanged.

For analytic semigroups we get another perturbation result:

Theorem 1.3.12. *Assume A generates an analytic semigroup. Then there exists a constant $\alpha > 0$ such that $(A + B, \mathcal{D}(A))$ generates an analytic semigroup for every A-bounded operator B with A-bound $a_0 < \alpha$.*

Proof. See [EN00, Theorem III.2.10]. □

1.3.2 H^∞-calculus and bounded imaginary powers

In the previous part we introduced C_0-semigroups as a way to solve the abstract Cauchy problem. As $T(t)u_0$ solves (1.4), the semigroup is often written as e^{tA}. We now want to generalise this concept and define functions of an operator A other than its exponential. We start by introducing the class of operators A we want to consider.

Definition 1.3.13. *A closed operator A on a Banach space X is called* sectorial *if*

a) *A is densely defined, injective and has dense range and*

b) *there exist $\varphi \in (0, \pi)$ and $C > 0$, such that $\sigma(A) \subseteq \overline{\Sigma_\varphi}$ and*

$$\|\lambda R(\lambda, A)\| \leq C \text{ for } \lambda \in \mathbb{C} \setminus \overline{\Sigma_\varphi}.$$

The infimum over all possible angles φ is called the sectoriality angle *of A and is denoted by φ_A.*

Remarks 1.3.14. 1. Many operators appearing in applications such as elliptic differential operators are sectorial, at least after multiplication by a scalar and translation.

2. If A is sectorial of angle $\varphi_A < \pi/2$, $-A$ is the generator of an analytic semigroup.

The next step is to introduce the functions we want to plug sectorial operators into. For $\theta \in (0, \pi]$ we define the algebra

$$H^\infty(\Sigma_\theta) = \{f : \Sigma_\theta \to \mathbb{C} : f \text{ is bounded and analytic}\}$$

and the subalgebra

$$H_0^\infty(\Sigma_\theta) = \left\{ f \in H^\infty(\Sigma_\theta) : \quad \text{there exist } s > 0, C > 0 \text{ such that} \right.$$
$$\left. |f(z)| \le C \frac{|z|^s}{1 + |z|^{2s}} \text{ for } z \in \Sigma_\theta \right\}.$$

Now let A be a sectorial operator with sectoriality angle φ_A, $\theta > \gamma > \varphi_A$, and set

$$\Gamma(t) = \begin{cases} te^{-i\gamma}, & t \ge 0, \\ -te^{i\gamma}, & t < 0. \end{cases}$$

Due to the resolvent estimate satisfied by A, the Dunford integral

$$f(A) := \frac{1}{2\pi i} \int_\Gamma f(\lambda) R(\lambda, A) d\lambda$$

exists for any $f \in H_0^\infty(\Sigma_\theta)$.

Definition 1.3.15. *A sectorial operator A on a Banach space X is said to admit a bounded H^∞-calculus if there exist an angle $\phi \in (\varphi_A, \pi)$ and a constant $C_\phi > 0$, such that*

$$(1.5) \qquad \|f(A)\|_{\mathcal{L}(X)} \le C_\phi \|f\|_\infty$$

for all $f \in H_0^\infty(\Sigma_\phi)$. The infimum over all angles for which an estimate (1.5) is valid is called the H^∞-angle of A. We denote the set of all sectorial operators admitting a bounded H^∞-calculus by $H^\infty(X)$.

If A admits a bounded H^∞-calculus, we can now define $f(A)$ for $f \in H^\infty(\Sigma_\theta)$ by the following procedure. Let $g(z) = z(1+z)^{-2}$. Then $g \in H_0^\infty(\Sigma_\theta)$ for any $\theta < \pi$ and using Cauchy's Theorem, we get a bounded and injective operator $g(A) = A(1+A)^{-2}$ with range $\text{rg}(g(A)) = D(A) \cap R(A)$. Now define

$$f(A) := (f \cdot g)(A) \, g(A)^{-1}$$

for $x \in D(A) \cap R(A)$. Using a density argument and the convergence lemma (cf. [CDMY96, Lemma 2.1]), we obtain a bounded linear operator $f(A)$ on X and the map $\Phi_A : f \mapsto f(A)$ is a bounded algebra homomorphism from $H^\infty(\Sigma_\theta)$ into $\mathcal{L}(X)$.

Of particular interest are the functions $f(z) = z^{is}$ for $s \in \mathbb{R}$. These are bounded holomorphic functions on the sectors Σ_θ for any $0 < \theta < \pi$. Using the above procedure, we can define A^{is} for sectorial operators A to obtain, in general, unbounded operators A^{is}. We introduce a new subclass of sectorial operators for which A^{is} is bounded.

Definition 1.3.16. *Let A be a sectorial operator on a Banach space X. We say that A has* bounded imaginary powers, *or $A \in \mathrm{BIP}(X)$, if $A^{is} \in \mathcal{L}(X)$ for all $s \in \mathbb{R}$ and there exists $C > 0$ such that*

$$\left\| A^{is} \right\|_{\mathcal{L}(X)} \leq C \ \text{for } |s| \leq 1.$$

For any sectorial operator, $(A^{is})_{s \in \mathbb{R}}$ forms a group (cf. [DHP03, Proposition 2.2]). Then $A \in \mathrm{BIP}(X)$, iff $(A^{is})_{s \in \mathbb{R}}$ is strongly continuous. The growth bound of this group

$$\theta_A = \limsup_{s \in \mathbb{R}} \frac{1}{|s|} \log \left\| A^{is} \right\|_{\mathcal{L}(X)}$$

is called the *power angle of A*.

A useful property of operators admitting bounded imaginary powers is that the domains of fractional powers of these operators are given by the complex interpolation method. In fact (cf. [Tri83, Theorem 1.15.3]),

$$D(A^\alpha) = [X, D(A)]_\alpha \ \text{for } \alpha \in (0, 1).$$

We now state some results that give sufficient conditions on A to admit bounded imaginary powers and allow estimates of the power angle θ_A. These results can be found in [Ama95, Example III.4.7.3]. In Hilbert spaces we have

Proposition 1.3.17. *Let A be a positive self-adjoint operator on a Hilbert space H. Then A has bounded imaginary powers and $\| A^{it} \|_{\mathcal{L}(H)} \leq 1$.*

The next propositions give bounds on the imaginary powers in L^p-spaces.

Proposition 1.3.18. *Let Ω be a domain in $\mathbb{R}^n$ and let $-A$ be the generator of a positive contraction semigroup of negative type on $L^p(\Omega)$ for all $1 < p < \infty$. Then*

$$\left\| A^{it} \right\|_{\mathcal{L}(L^p(\Omega))} \leq C_p(1 + |t|)e^{|t|\pi/2} \text{ for } t \in \mathbb{R}.$$

By interpolation of the previous propositions, we obtain

Proposition 1.3.19. *Let Ω be a domain in $\mathbb{R}^n$ and let $-A$ be the generator of a positive contraction semigroup of negative type on $L^p(\Omega)$ for all $1 < p < \infty$. Moreover, let A be self-adjoint in $L^2(\Omega)$. Then A has bounded imaginary powers of angle $\theta < \pi/2$ on $L^p(\Omega)$ for all $1 < p < \infty$.*

1.3.3 Maximal regularity

An interesting property operators A can have is maximal regularity. This property guarantees existence and uniqueness of solutions to a parabolic problem involving A in certain function spaces.

Definition 1.3.20. *Let X be a Banach space and A a densely defined closed linear operator on X. Consider the parabolic problem*

$$(1.6) \qquad \begin{cases} u'(t) - Au(t) = f(t) & \text{for } t \in \mathbb{R}_+, \\ \quad u(0) = 0. \end{cases}$$

We say A has maximal L^q-regularity in X, written $A \in MR(q, X)$, if for every $f \in L^q(\mathbb{R}_+, X)$ there exists a unique solution to (1.6) such that $u' \in L^q(\mathbb{R}_+, X)$.

Remarks 1.3.21. 1. Often the additional requirement that $u \in L^q(\mathbb{R}_+, X)$ is made. However, this requires that the spectral bound $s(A) = \sup\{\text{Re } \lambda : \lambda \in \sigma(A)\} < 0$ (cf. [Dor93]). Since our main focus is on the Laplacian which we cannot expect to satisfy this condition in unbounded domains, we will work with the weaker definition.

2. If $A \in MR(q, X)$, we have that $Au \in L^q(\mathbb{R}_+, X)$ for the solution u of (1.6) and from the closed graph theorem we get the important estimate

$$\|u'\|_{L^q(\mathbb{R}_+, X)} + \|Au\|_{L^q(\mathbb{R}_+, X)} \leq C \|f\|_{L^q(\mathbb{R}_+, X)}.$$

If u itself lies in $L^q(\mathbb{R}_+, X)$, we even get

$$\|u\|_{L^q(\mathbb{R}_+, X)} + \|u'\|_{L^q(\mathbb{R}_+, X)} + \|Au\|_{L^q(\mathbb{R}_+, X)} \leq C \|f\|_{L^q(\mathbb{R}_+, X)}.$$

3. Maximal regularity in X for one $q \in (1, \infty)$ implies maximal regularity in X for all $q \in (1, \infty)$ (see [Dor93]).

4. Maximal regularity of A implies that A generates a bounded analytic C_0-semigroup (cf. [Dor93] or [HP97]).

5. In Hilbert spaces, maximal regularity for A is in fact equivalent to A being the generator of a bounded analytic C_0-semigroup (cf. [dS64]). However, this equivalence is no longer true in Banach spaces.

6. If X is a complex Banach space and A has maximal L^q-regularity in X, then A has maximal L^q-regularity in X on $[0, T]$ for all $T \in \mathbb{R}_+$, i.e. we can replace $\mathbb{R}_+$ by the interval $[0, T]$ in Definition 1.3.20 (cf. [Dor93]).

We now state two theorems giving sufficient conditions for A to have maximal regularity. The Dore-Venni theorem gives maximal regularity of an operator admitting bounded imaginary powers with a sufficiently small power angle.

Theorem 1.3.22 (Dore, Venni). *Let Ω be a domain in $\mathbb{R}^n$ and X be the Banach space $L^p(\Omega)$ for some $1 < p < \infty$.[5] Let $A : D(A) \to X$ be a densely defined linear operator in X which satisfies the resolvent estimate*

$$\left\|(\lambda - A)^{-1}\right\| \leq \frac{C}{1 + \lambda} \text{ for all } \lambda > 0,$$

and such that $-A$ has bounded imaginary powers of angle $\theta_{-A} < \pi/2$. Then A has maximal L^q-regularity on X for $1 < q < \infty$. Furthermore, the solution u to the parabolic problem (1.6) lies in $L^q(\mathbb{R}_+, X)$ and we have the estimate

$$\|u\|_{L^q(\mathbb{R}_+, X)} + \|u'\|_{L^q(\mathbb{R}_+, X)} + \|Au\|_{L^q(\mathbb{R}_+, X)} \leq C \|f\|_{L^q(\mathbb{R}_+, X)}.$$

---

[5]In fact, the theorem holds for all UMD-spaces X.

Proof. Cf. [DV87]. □

Instead of using bounded imaginary powers of the operator, we can prove maximal regularity of an operator via $\mathcal{R}$-boundedness [6]. The relevant result for us is the following (cf. [Wei01b, Corollary 4d]):

Theorem 1.3.23 (Weis)**.** *Let Ω be a domain in $\mathbb{R}^n$. If A generates a positive analytic semigroup of contractions on $L^p(\Omega)$ for some $1 < p < \infty$, then A has maximal L^q-regularity on $L^p(\Omega)$ for all $1 < q < \infty$.*

1.3.4 Fourier multipliers

For a function $m \in L^\infty(\mathbb{R}^n, \mathbb{C})$, we consider the linear operator T defined by

$$Tf(x) = \left(\mathcal{F}^{-1}(m \cdot \mathcal{F}f) \right)(x)$$

as a map from $\mathcal{S}(\mathbb{R}^n)$ into $\mathcal{S}'(\mathbb{R}^n)$. If T extends to a bounded operator on $L^p(\mathbb{R}^n)$, $1 \leq p < \infty$, we call m a *Fourier multiplier* on $L^p(\mathbb{R}^n)$, and we denote the set of Fourier multipliers by $M_p(\mathbb{R}^n)$.

If $p = 2$, by Plancherel's Theorem we obtain that

$$\|Tf\|_2 = \|m \cdot \mathcal{F}f\|_2 \leq \|m\|_\infty \|f\|_2,$$

so $M_2(\mathbb{R}^n) = L^\infty(\mathbb{R}^n)$. For other p however, the situation is more difficult. A sufficient condition for m to be a Fourier multiplier on $L^p(\mathbb{R}^n)$ is given by Mikhlin's Theorem.

Theorem 1.3.24 (Mikhlin)**.** *Let $m \in C^k(\mathbb{R}^n \setminus \{0\})$ for $k = \min\{j \in \mathbb{N} : j > \frac{n}{2}\}$ and suppose*

$$\|m\|_M := \max_{|\alpha| \leq k} \sup_{\xi \in \mathbb{R}^n \setminus \{0\}} |\xi|^{|\alpha|} |D^\alpha m(\xi)| < \infty.$$

Then $m \in M_p(\mathbb{R}^n)$ for all $1 < p < \infty$ and there exists a constant $C > 0$, such that the operator $T = \mathcal{F}^{-1}m\mathcal{F}$ satisfies the estimate $\|T\|_{\mathcal{L}(L^p(\mathbb{R}^n))} \leq C \|m\|_M$.

Corollary 1.3.25. *For $k \in \mathbb{N}$ and $1 < p < \infty$ we have $W^{k,p}(\mathbb{R}^n) = L_k^p(\mathbb{R}^n)$.*

For more results on Fourier multipliers and a proof of Mikhlin's Theorem, we refer to [Ste93, Chapter VI].

---

[6]See [Wei01a], [Wei01b] and [DHP03] for more on the concept of $\mathcal{R}$-boundedness.

Chapter 2

Laplace's equation in Lipschitz domains

In this chapter we consider Laplace's equation

$$(2.1) \qquad \Delta u = f \quad \text{in } \Omega$$

for data f in some Sobolev space $W^{\alpha,p}(\Omega)$, possibly of negative order. The domain Ω is assumed to be Lipschitz and we consider (2.1) for Dirichlet and for Neumann boundary conditions.

In the first section, we state the main results known for bounded domains. Theorem 2.1.1 is due to Jerison and Kenig and deals with the case of Dirichlet boundary conditions. Theorem 2.1.6, due to Fabes, Mendez and Mitrea, treats the Neumann problem. Unlike in the case of smooth domains, where the solution u gains two degrees of regularity over the data regardless of the exponent p, in Lipschitz domains we need to restrict the range of α and p to obtain this result. By making the additional assumption that Ω is convex, we can extend the range of α and p substantially. This is done in Theorem 2.1.2 by Fromm for the Dirichlet problem and in Theorem 2.1.8 by Adolfsson and Jerison in the Neumann case. We also mention an interesting result by Shen on the resolvent problem (Theorem 2.1.10) which gives first results for the parabolic problem.

In the second section we extend these results for unbounded domains Ω by approximating them by bounded domains. This leads to solutions in the homogeneous Sobolev spaces introduced in Section 1.2.5. For convex domains Ω and data $f \in L^p(\Omega)$, this leads to a slight improvement on a result by Adolfsson (Theorem 2.1.5).

2.1   Known results for bounded domains

We first state the main result due to Jerison and Kenig on solutions to the Dirichlet problem in bounded Lipschitz domains.

Theorem 2.1.1. (Jerison, Kenig [JK95, Theorem 1.1]). *Let Ω be a bounded Lipschitz domain in $\mathbb{R}^n$, $n \geq 3$. There exists $\varepsilon \in (0,1]$, depending only on the Lipschitz character[1] of Ω such that for every $f \in L^p_{\alpha-2}(\mathbb{R}^n)$ there is a unique solution $u \in L^p_\alpha(\Omega)$ to the inhomogeneous Dirichlet problem*

$$(2.2) \qquad \begin{cases} \Delta u = f & in\ \Omega, \\ u = 0 & on\ \partial\Omega, \end{cases}$$

provided one of the following holds:

1. $1 < p \leq p_0$ and $\dfrac{3}{p} - 1 - \varepsilon < \alpha < 1 + \dfrac{1}{p}$,

2. $p_0 < p < p'_0$ and $\dfrac{1}{p} < \alpha < 1 + \dfrac{1}{p}$,

3. $p'_0 \leq p < \infty$ and $\dfrac{1}{p} < \alpha < \dfrac{3}{p} + \varepsilon$,

where $1/p_0 = 1/2 + \varepsilon/2$ and $1/p'_0 = 1/2 - \varepsilon/2$. Moreover, we have the estimate

$$\|u\|_{L^p_\alpha(\Omega)} \leq C \|f\|_{L^p_{\alpha-2}(\mathbb{R}^n)}$$

for all $f \in L^p_{\alpha-2}(\mathbb{R}^n)$. The constant C depends on the domain Ω only via the Lipschitz character of Ω. When the domain is C^1, the exponent p_0 may be taken to be 1.

---

[1]That is the number of coordinate charts used to cover the boundary $\partial\Omega$ by cylinders such that, inside each cylinder, the domain is the domain above the graph of a Lipschitz function, the radii of these cylinders and the supremum of the norms of the graph functions.

Note that in the theorem $\alpha < 2$, so it does not give $W^{2,p}$-solutions for any p. As early as 1979 Dahlberg [Dah79] constructed a bounded Lipschitz domain $\Omega \subseteq \mathbb{R}^2$ and data $f \in L^\infty(\Omega)$ such that the solution to (2.2) is not in $W^{2,p}(\Omega)$ for any $1 < p < \infty$. In [JK95], Jerison and Kenig are able to improve this result to a C^1-domain. Moreover, they show that the range of α and p in Theorem 2.1.1 is optimal for Lipschitz domains.

For convex bounded domains $\Omega \subseteq \mathbb{R}^n$ and $1 < p \leq 2$, it is possible to control all second derivatives in $L^p(\Omega)$ by the Laplacian. Note that any bounded convex domain is a Lipschitz domain (cf. [Gri85, Corollary 1.2.2.3]). The following result is due to Fromm.

Theorem 2.1.2 (Fromm [Fro93]). *If $\Omega \subset \mathbb{R}^n$, $n \geq 2$, is a bounded and convex domain with diameter d and if $f \in L_s^p(\Omega)$ then there is a unique $u \in W_0^{1,p}(\Omega) \cap L_{s+2}^p(\Omega)$ satisfying $\Delta u = f$ in Ω. This solution satisfies the estimate*

$$(2.3) \qquad \|u\|_{s+2,p} \leq C(d) \, \|f\|_{s,p}$$

for $-1 \leq s \leq 0$ and $1 < p < \frac{2}{s+1}$ (defining $\frac{2}{0} = \infty$) and for $s = 0$, $p = 2$.

The theorem actually holds in a larger class of domains which are Lipschitz domains that are convex in the neighbourhood of any boundary singularities.

Definition 2.1.3. *Let Ω be a domain in $\mathbb{R}^n$. We say that Ω satisfies the outer ball condition if for each $x \in \partial\Omega$, there exists an open ball $B \subseteq \Omega^c$ with $x \in \partial B$. Ω satisfies a uniform outer ball condition if there exists an $R > 0$ such that for all $x \in \partial\Omega$, the ball can be chosen to have radius R.*

Remarks 2.1.4. 1. The outer ball condition guarantees that wherever the boundary of Ω is not of class C^1, the singularity is directed outwards.

2. Theorem 2.1.2 holds in all bounded Lipschitz domains satisfying a uniform outer ball condition (cf. [Fro93, Remarks]). In this case, the constant C in (2.3) depends on more geometric properties of Ω than just the diameter.

It turns out that we can drop the requirement that Ω be bounded if we only want the second derivatives to be in $L^p(\Omega)$ and the domain is a convex domain above a Lipschitz graph. This result was first proved by Adolfsson.

Theorem 2.1.5 (Adolfsson [Ado93]). *Let* $\Omega \subseteq \mathbb{R}^n$, $n \geq 3$, *be a convex domain above a Lipschitz graph given by a function* φ *with Lipschitz constant* L. *Suppose* $f \in L^p(\Omega)$ *and* $1 < p \leq 2$. *Then there is a solution* u *to the Dirichlet problem (2.2) given by the Green potential of* f *and it satisfies*

$$\int_\Omega |\nabla^2 u|^p \leq C \int_\Omega |f|^p,$$

where ∇^2 *denotes the matrix of the second order derivatives and* C *only depends on the Lipschitz constant* L.

Theorem 2.1.2 is optimal in the sense that for any $1 < p < \infty$ with $p > \frac{2}{s+1}$ and $-1 < s < 1$, there exists a bounded convex domain Ω and a function $f \in C_c^\infty(\overline{\Omega})$ where the solution to the Dirichlet problem is not in $L_{s+2}^p(\Omega)$. In particular, even in bounded convex domains, the solution to

$$\begin{cases} \Delta u = f & \text{in } \Omega, \\ \quad u = 0 & \text{on } \partial\Omega \end{cases}$$

is not necessarily in $W^{2,p}(\Omega)$ for any $p > 2$ (cf. [Fro93]).

The results corresponding to Theorem 2.1.1 and Theorem 2.1.5 for Neumann boundary conditions are due to Fabes, Mendez and Mitrea [FMM98] and, in the convex case, Adolfsson and Jerison [AJ94]. They read as follows.

Theorem 2.1.6 (Fabes, Mendez, Mitrea [FMM98]). *Suppose* Ω *is a bounded Lipschitz domain in* $\mathbb{R}^n$, $n \geq 3$. *Then there exists* $\varepsilon \in (0, 1]$, *such that for every* $f \in L_{\alpha-2,0}^p(\Omega)$ *satisfying the compatibility condition* $\langle f, \chi_\Omega \rangle = 0$ *there is a solution* $u \in L_\alpha^p(\Omega)$ *to the inhomogeneous Neumann problem*

(2.4)
$$\begin{cases} \Delta u = f & \text{in } \Omega, \\ \frac{\partial u}{\partial N} = 0 & \text{on } \partial\Omega, \end{cases}$$

provided one of the following holds:

1. $1 < p \leq p_0$ *and* $\dfrac{3}{p} - 1 - \varepsilon < \alpha < 1 + \dfrac{1}{p}$,

2. $p_0 < p < p_0'$ *and* $\dfrac{1}{p} < \alpha < 1 + \dfrac{1}{p}$,

3. $p_0' \leq p < \infty$ *and* $\dfrac{1}{p} < \alpha < \dfrac{3}{p} + \varepsilon$,

where $1/p_0 = 1/2 + \varepsilon/2$ and $1/p_0' = 1/2 - \varepsilon/2$. Moreover, the solution is unique up to a constant and we have the estimate

$$\inf_{c \in \mathbb{C}} \|u - c\|_{L_\alpha^p(\Omega)} \leq C \|f\|_{L_{\alpha-2}^p(\mathbb{R}^n)}.$$

When the domain Ω is C^1, the exponent p_0 may be taken to be 1.

Remarks 2.1.7. 1. Note that the range of α and p for which the theorem holds is the same as given in Theorem 2.1.1 for the Dirichlet problem.

2. In fact, in [FMM98] inhomogeneous boundary conditions are considered and the corresponding problems are solved for boundary data in suitable Besov spaces both for the Dirichlet and for the Neumann problem.

As in the case of Dirichlet boundary conditions, for bounded convex domains we can also control the second derivatives.

Theorem 2.1.8 (Adolfsson, Jerison [AJ94]). *Let $\Omega \subseteq \mathbb{R}^n$ be a bounded convex domain and $n \geq 3$. Suppose $f \in L^p(\Omega)$ with $\int_\Omega f = 0$ and $1 < p \leq 2$. Then there exists a solution u to the Neumann problem (2.4) in $W^{2,p}(\Omega)$ and*

$$\int_\Omega |\nabla^2 u|^p \leq C \int_\Omega |f|^p,$$

where ∇^2 denotes the matrix of the second order derivatives and C only depends on the Lipschitz constant of the domain.

Remarks 2.1.9. Note that by Theorem 2.1.6, the solution is unique up to a constant.

Returning to Dirichlet boundary conditions, Shen [She95] considers the resolvent problem for elliptic systems with real constant coefficients. In particular, for the Laplacian in bounded Lipschitz domains in $\mathbb{R}^n$, $n \geq 3$, he shows resolvent estimates which guarantee that the Laplacian generates a bounded analytic C_0-semigroup and he estimates the gradient of the resolvent. For the Laplacian his results read as follows.

Theorem 2.1.10 (Shen [She95]). *Let Ω be a bounded Lipschitz domain in $\mathbb{R}^n$ with connected boundary and $\lambda \in \Sigma_\theta$ for some $\theta < \pi$. Let $f \in L^p(\Omega)$.*

Consider the problem

(2.5)
$$\begin{cases} (\lambda - \Delta)u = f & in\ \Omega, \\ \qquad\quad u = 0 & on\ \partial\Omega. \end{cases}$$

Then

a) *if $n \geq 4$, there exist $C > 0$, $\delta > 0$, depending on Ω, n and θ, such that if $2n/(n+3) - \delta < p < 2n/(n-3) + \delta$, then there is a unique solution u of (2.5) and it satisfies*

(2.6)
$$\|u\|_p \leq \frac{C}{1 + |\lambda|} \|f\|_p,$$

b) *if $n = 3$ and $1 \leq p \leq \infty$, then there is a unique solution u of (2.5) and it satisfies (2.6).*

Moreover, for $n \geq 3$, we have the gradient estimate

(2.7)
$$\|\nabla u\|_p \leq \frac{C}{(1 + |\lambda|)^{\frac{1}{2}}} \|f\|_p$$

for $2n/(n+1) - \delta < p < 2n/(n-1) + \delta$.

2.2   Generalisations for unbounded domains

Our aim in this section is to extend the theorems by Jerison and Kenig (Theorem 2.1.1) and Fromm (Theorem 2.1.2) for Laplace's equation to unbounded Lipschitz domains. We will see that this causes difficulties. One reason for this is that functions with derivatives in $L^p(\Omega)$ are not necessarily L^p-functions themselves. This forces us to consider problems in homogeneous Sobolev spaces.

2.2.1   Approximation by bounded domains

The first step is to determine the dependence of the constants in estimates such as (2.3) for solutions of Laplace's equation in bounded domains on the geometry of the domain.

Let Ω be either the domain over the graph of a Lipschitz function φ or a convex domain.

Let $f \in C_c^\infty(\Omega)$ and $\Omega_R := \Omega \cap C_R$, where C_R is a cylinder centered at the origin of height $2R$ with axis in the graph direction, and with a base of diameter R with R large enough such that f is supported in Ω_R. Consider

$$(2.8) \qquad \begin{cases} \Delta u_R = f & \text{in } \Omega_R, \\ u_R = 0 & \text{on } \partial\Omega_R. \end{cases}$$

Further let $\Omega_1^R = \{x \in \mathbb{R}^n : Rx \in \Omega_R\}$, $f_1(x) = f(Rx)$ and $u_1(x) = u_R(Rx)$ for $x \in \Omega_1^R$. Then u_1 solves

$$\begin{cases} \Delta u_1 = R^2 f_1 & \text{in } \Omega_1^R, \\ u_1 = 0 & \text{on } \partial\Omega_1^R \end{cases}$$

and by Theorem 2.1.1 in the case of a Lipschitz graph or by Theorem 2.1.2 in the case of a convex domain, we have that

$$(2.9) \qquad \|u_1\|_{W^{1,p}(\Omega_1^R)} \le CR^2 \|f_1\|_{W^{-1,p}(\Omega_1^R)}$$

for $1 < p < 3 + \varepsilon$. In the convex case, the estimate holds for any $1 < p < \infty$.

Remarks 2.2.1. The constant C in estimate (2.9) is independent of R. This is true since, for Lipschitz graphs, the constant in Theorem 2.1.1 depends on a domain D only via the Lipschitz character of D. As the boundary of Ω is described by a Lipschitz function and rescaling does not affect the Lipschitz character, the Lipschitz character of the domains Ω_1^R is uniformly bounded in R. For convex domains, Theorem 2.1.2 states that the constant only depends on the diameter of the domain. However, $\Omega_1^R \subseteq B_{\sqrt{2}}$, so again the constant in estimate (2.9) is independent of R.

At this point it is important to note, as we did in Chapter 1, that although $W^{1,p}(\Omega) = L_1^p(\Omega)$ as sets, the norms can differ. As some of the results we use from Section 2.1 are formulated in the Bessel potential spaces, we need to show that for $u \in W_0^{1,p}(\Omega)$ there exists a constant C independent of Ω such that

$$\frac{1}{C} \left\| (1-\Delta)^{\frac{1}{2}} u \right\|_{L^p(\Omega)} \le \|u\|_{L^p(\Omega)} + \|\nabla u\|_{L^p(\Omega)} \le C \left\| (1-\Delta)^{\frac{1}{2}} u \right\|_{L^p(\Omega)}.$$

To do this, first consider the whole space case. Let $u \in W^{1,p}(\mathbb{R}^n)$. Then

$$
\begin{aligned}
\|u\|_{L^p(\mathbb{R}^n)} &= \left\|\mathcal{F}^{-1}(\mathcal{F}u)(\cdot)\right\|_{L^p(\mathbb{R}^n)}, \\
\|\nabla u\|_{L^p(\mathbb{R}^n)} &= \left\|\mathcal{F}^{-1}(\cdot \mathcal{F}u(\cdot))\right\|_{L^p(\mathbb{R}^n)}, \\
\left\|(1-\Delta)^{\frac{1}{2}}u\right\|_{L^p(\mathbb{R}^n)} &= \left\|\mathcal{F}^{-1}\left((1+|\cdot|^2)^{\frac{1}{2}}\mathcal{F}u(\cdot)\right)\right\|_{L^p(\mathbb{R}^n)}.
\end{aligned}
$$

Now, by Mikhlin's multiplier theorem (Theorem 1.3.24),

$$
\begin{aligned}
m_1(\xi) &= \frac{1}{(1+|\xi|^2)^{\frac{1}{2}}}, \\
m_2(\xi) &= \frac{i\xi}{(1+|\xi|^2)^{\frac{1}{2}}} \quad \text{and} \\
m_{3,i}(\xi) &= \frac{\xi_i}{(1+|\xi|^2)^{\frac{1}{2}}}, \quad 1 \le i \le n,
\end{aligned}
$$

are all Fourier multipliers, so we obtain that for all $u \in W^{1,p}(\mathbb{R}^n)$,

$$
\|u\|_{L^p(\mathbb{R}^n)} + \|\nabla u\|_{L^p(\mathbb{R}^n)} \le \left(\|m_1\|_M + \|m_2\|_M\right)\left\|(1-\Delta)^{\frac{1}{2}}u\right\|_{L^p(\mathbb{R}^n)}.
$$

On the other hand,

$$
(1+|\xi|^2)^{\frac{1}{2}} = \frac{1}{(1+|\xi|^2)^{\frac{1}{2}}} + \sum_{i=1}^{n}\frac{\xi_i}{(1+|\xi|^2)^{\frac{1}{2}}}\xi_i
$$

implies that

$$
\left\|(1-\Delta)^{\frac{1}{2}}u\right\|_{L^p(\mathbb{R}^n)} \le \|m_1\|_M\|u\|_{L^p(\mathbb{R}^n)} + \sum_{i=1}^{n}\|m_{3,i}\|_M\|\nabla u\|_{L^p(\mathbb{R}^n)}.
$$

Next consider $u \in C_c^\infty(\Omega)$. Extend u trivially to $\mathbb{R}^n$. Then

$$
\|u\|_{L^p(\Omega)} = \|u\|_{L^p(\mathbb{R}^n)}, \quad \|\nabla u\|_{L^p(\Omega)} = \|\nabla u\|_{L^p(\mathbb{R}^n)}
$$

and by the definition of the Bessel potential spaces on domains, we have

$$
\left\|(1-\Delta)^{\frac{1}{2}}u\right\|_{L^p(\Omega)} \le \left\|(1-\Delta)^{\frac{1}{2}}u\right\|_{L^p(\mathbb{R}^n)}.
$$

Therefore, we find a constant independent of Ω such that

$$
\left\|(1-\Delta)^{\frac{1}{2}}u\right\|_{L^p(\Omega)} \le C(\|u\|_{L^p(\Omega)} + \|\nabla u\|_{L^p(\Omega)}).
$$

On the other hand, let v be any extension of u in $W^{1,p}(\mathbb{R}^n)$. Then

$$
\begin{aligned}
\|u\|_{L^p(\Omega)} + \|\nabla u\|_{L^p(\Omega)} &\leq \|v\|_{L^p(\mathbb{R}^n)} + \|\nabla v\|_{L^p(\mathbb{R}^n)} \\
&\leq C \left\| (1-\Delta)^{\frac{1}{2}} v \right\|_{L^p(\mathbb{R}^n)}.
\end{aligned}
$$

Taking the infimum on the right hand side over all extensions v we get that

$$
\|u\|_{L^p(\Omega)} + \|\nabla u\|_{L^p(\Omega)} \leq C \left\| (1-\Delta)^{\frac{1}{2}} u \right\|_{L^p(\Omega)}.
$$

By a density argument, the equivalence of norms with constant independent of Ω then holds for $u \in W_0^{1,p}(\Omega)$.

Recalling the homogeneous Sobolev spaces introduced in Section 1.2.5, we are now able to determine the R-dependence of the constant C_R in the estimate

$$
\|u_R\|_{W^{1,p}(\Omega_R)} \leq C_R \|f\|_{W^{-1,p}(\Omega_R)}
$$

for the solution to problem (2.8) whenever we choose the data $f \in C_c^{\infty}(\Omega) \cap \widehat{\mathcal{W}}^{-1,p}(\Omega)$.

Proposition 2.2.2. *For $f \in C_c^{\infty}(\Omega) \cap \widehat{\mathcal{W}}^{-1,p}(\Omega)$, the solution to (2.8) satisfies the following estimate*

$$
(2.10) \qquad \|u_R\|_{L^p(\Omega_R)} + R \|\nabla u_R\|_{L^p(\Omega_R)} \leq CR \|f\|_{\widehat{\mathcal{W}}^{-1,p}(\Omega_R)},
$$

where the constant C is independent of R.

Proof. A calculation shows that

$$
\begin{aligned}
\|u_1\|_{L^p(\Omega_1^R)}^p &= R^{-n} \|u_R\|_{L^p(\Omega)}^p, \\
\|\partial_i u_1\|_{L^p(\Omega_1^R)}^p &= R^{-n+p} \|\partial_i u_R\|_{L^p(\Omega)}^p.
\end{aligned}
$$

Furthermore,

$$
\begin{aligned}
&\|f_1\|_{W^{-1,p}(\Omega_1^R)} \\
&= \sup \left\{ \int_{\Omega_1^R} f(Rx)h(x)\mathrm{d}x : h \in C_c^{\infty}(\Omega_1^R), \|h\|_{W^{1,p'}(\Omega_1^R)} = 1 \right\} \\
&= R^{-n} \sup \left\{ \int_{\Omega_R} f(y)g(y)\mathrm{d}y : g \in C_c^{\infty}(\Omega_R), \right.
\end{aligned}
$$

$$R^{-\frac{n}{p'}} \|g\|_{L^{p'}(\Omega_R)} + R^{1-\frac{n}{p'}} \|\nabla g\|_{L^{p'}(\Omega_R)} = 1 \bigg\}$$

$$= R^{-n-1+\frac{n}{p'}} \sup \bigg\{ \int_{\Omega_R} f(y)g(y)\mathrm{d}y : g \in C_c^\infty(\Omega_R),$$

$$R^{-1} \|g\|_{L^{p'}(\Omega_R)} + \|\nabla g\|_{L^{p'}(\Omega_R)} = 1 \bigg\}$$

$$\leq R^{-1-\frac{n}{p}} \sup \bigg\{ \int_{\Omega_R} f(y)g(y)\mathrm{d}y : g \in C_c^\infty(\Omega_R), \|\nabla g\|_{L^{p'}(\Omega_R)} = 1 \bigg\}$$

$$\leq R^{-1-\frac{n}{p}} \sup \bigg\{ \int_{\Omega_R} f(y)g(y)\mathrm{d}y : g \in \widehat{\mathcal{W}}_0^{1,p}(\Omega_R), \|\nabla g\|_{L^{p'}(\Omega_R)} = 1 \bigg\}$$

$$= R^{-1-\frac{n}{p}} \|f\|_{-1,p}^\wedge .$$

Combining these estimates with (2.9), we obtain (2.10). $\qquad\qquad\square$

2.2.2   The solution in unbounded domains

Having determined the dependence of solutions of the Dirichlet problem in bounded domains on the geometry of the domain, we can now use these results to obtain solutions in unbounded domains.

Assume that $f \in C_c^\infty(\Omega) \cap \widehat{\mathcal{W}}^{-1,p}(\Omega)$. By (2.10), we find that there is a subsequence of the u_R such that for all $g \in \big(L^{p'}(\Omega) \big)^n$ we have

$$(2.11) \qquad \int_\Omega \nabla u_R \cdot g \to \int_\Omega v \cdot g \text{ as } R \to \infty$$

for some $v \in (L^p(\Omega))^n$.

We next show that the functions u_R are locally uniformly bounded in $L^p(\Omega)$. Let Ω' be a bounded domain in $\mathbb{R}^{n-1}$, $M > \|\varphi\|_\infty$, where φ is the Lipschitz function describing the boundary of Ω and $\widetilde{\Omega} = \Omega' \times (-M, M)$. See Figure 2.1. Extend u_R to those parts of $\widetilde{\Omega}$ outside Ω_R by zero. By Poincaré's inequality (cf. Theorem 1.2.15), we get $\|u_R\|_{L^p(\widetilde{\Omega})} \leq C_M \|\nabla u_R\|_{L^p(\widetilde{\Omega})}$.[2]

As any bounded domain $\Omega_b \subseteq \Omega$ is contained in a domain of the form $\widetilde{\Omega}$, we now know that for any bounded $\Omega_b \subseteq \Omega$ there exists a subsequence (u_R) and

---

[2] Actually, we use a slightly stronger version of Poincaré's inequality for which it suffices that u_R vanishes on the lower part of the boundary of $\widetilde{\Omega}$.

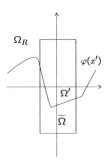

Figure 2.1: The domain $\widetilde{\Omega}$.

$u \in L^p(\Omega_b)$ such that for all $g \in \left(L^{p'}(\Omega) \right)^n$ we have

$$\int_\Omega \nabla u_R \cdot g \to \int_\Omega v \cdot g \ \text{ as } R \to \infty$$

and for all $h \in L^{p'}(\Omega_b)$,

$$\int_{\Omega_b} u_R h \to \int_{\Omega_b} uh \ \text{ as } R \to \infty.$$

Let $g \in C_c^\infty(\Omega_b)^n$ and $h = \text{div } g$. Then $h \in C_c^\infty(\Omega_b)$ and we have

$$
\begin{aligned}
\int_{\Omega_b} (v - \nabla u) \cdot g &= \lim_{R \to \infty} \left(\int_{\Omega_b} \nabla u_R \cdot g + \int_{\Omega_b} u_R h \right) \\
&= \lim_{R \to \infty} \int_{\Omega_b} (-u_R h + u_R h) = 0.
\end{aligned}
$$

Therefore, $v = \nabla u$ in $L^p(\Omega_b)$ and we can find $u \in L^p_{\text{loc}}(\Omega)$ such that $\nabla u = v$ in $L^p(\Omega)$.

Theorem 2.2.3. *Let Ω be the domain over the graph of a Lipschitz function φ and $1 < p < 3 + \varepsilon$ or let Ω be a convex domain and $1 < p < \infty$. Given $f \in \widehat{\mathcal{W}}^{-1,p}(\Omega)$, there is a solution $u \in \widehat{W}_0^{1,p}(\Omega)$ to*

$$
\begin{cases}
\Delta u = f & in \ \Omega, \\
u = 0 & on \ \partial\Omega
\end{cases}
$$

and ∇u satisfies the estimate

$$(2.12) \qquad\qquad \|\nabla u\|_p \leq C \|f\|^{\wedge}_{-1,p} .$$

Proof. First take a function $f \in C_c^\infty(\Omega) \cap \widehat{\mathcal{W}}^{-1,p}(\Omega)$ and let u_R be the solution of (2.8). Then, let u be the L^p_{loc}-function in $\widehat{\mathcal{W}}^{1,p}(\Omega)$ gained from the functions u_R as above, and suppose $\phi \in C_c^\infty(\Omega)$. Then $\nabla \phi \in \left(L^{p'}(\Omega)\right)^n$. Inserting this in (2.11) and integration by parts yields

$$\int_\Omega \Delta u \phi = - \int_\Omega \nabla u \nabla \phi = - \lim_{R\to\infty} \int_\Omega \nabla u_R \nabla \phi = \lim_{R\to\infty} \int_\Omega \Delta u_R \phi = \int_\Omega f\phi,$$

i.e. $\Delta u = f$ in $\mathcal{D}'(\Omega)$.

Since the trace operator $\widetilde{\mathrm{tr}} : \widehat{\mathcal{W}}^{1,p}(\Omega) \to \widetilde{\mathrm{tr}}(\widehat{\mathcal{W}}^{1,p}(\Omega))$ is continuous (cf. Section 1.2.5), it is weakly continuous, so

$$0 = \widetilde{\mathrm{tr}}\, u_R \rightharpoonup \widetilde{\mathrm{tr}}\, u.$$

This yields $\widetilde{\mathrm{tr}}\, u = 0$ by the uniqueness of the weak limit (cf. [RR93, p.203]), i.e. tr u is constant and we can choose the representative u so that the boundary condition is satisfied.

We already know that ∇u_R converges weakly to ∇u in $L^p(\Omega)$ and satisfies the estimate $\|\nabla u_R\|_p \leq C \|f\|^{\wedge}_{-1,p}$. Then (2.12) follows from a standard result on weak convergence stating that $\|f\| \leq \liminf_{n\in\mathbb{N}} \|f_n\|$ whenever f_n converges weakly to f (cf. [Alt85, Bemerkung 6.3] or [Tay67, Section 4.41]).

For the general case, when $f \in \widehat{\mathcal{W}}^{-1,p}(\Omega)$, approximate f in $\widehat{\mathcal{W}}^{-1,p}(\Omega)$ by $f_n \in C_c^\infty(\Omega) \cap \widehat{\mathcal{W}}^{-1,p}(\Omega)$. Then solve

$$\begin{cases} \Delta u_n = f_n & \text{in } \Omega, \\ u_n = 0 & \text{on } \partial\Omega. \end{cases}$$

We have

$$\|\nabla u_n\|_p \leq C \|f_n\|^{\wedge}_{-1,p} \leq C.$$

Therefore, there exists $v \in L^p(\Omega)$ such that $\nabla u_n \rightharpoonup v$ in $L^p(\Omega)$. Locally, we have Poincaré's inequality, so for any bounded $\Omega_b \subseteq \Omega$ such that $\partial\Omega_b \cap \partial\Omega \neq \emptyset$, we have

$$\|u_n\|_{L^p(\Omega_b)} \leq C \|\nabla u_n\|_{L^p(\Omega_b)} \leq C$$

and there exists a function $u \in L^p(\Omega_b)$ such that $u_n \rightharpoonup u$ in $L^p(\Omega_b)$. Then integration by parts yields $v = \nabla u$ and $\Delta u = f$. As in the smooth case, using the continuity of $\widetilde{\text{tr}}$, we see that u also satisfies the boundary condition and that (2.12) holds. □

2.2.3 Convex domains

Using a similar procedure as for the first derivatives, we can use Fromm's result (Theorem 2.1.2) to prove boundedness of all second derivatives in $L^p(\Omega)$ for $1 < p \leq 2$ and data $f \in L^p(\Omega)$ when Ω is a convex domain. This gives an alternative proof of Adolfsson's result (Theorem 2.1.5) with the slight improvement that we actually obtain a constant C independent of Ω in the estimate (2.13).

Theorem 2.2.4. *Let Ω be a convex domain in $\mathbb{R}^n$, $n \geq 3$, and $1 < p \leq 2$. Then there is a solution $u \in L^p_{\text{loc}}(\Omega)$ to*

$$\begin{cases} \Delta u = f & \text{in } \Omega, \\ u = 0 & \text{on } \partial\Omega \end{cases}$$

with $f \in L^p(\Omega)$ and

$$(2.13) \qquad \left\|\nabla^2 u\right\|_p \leq C \left\|f\right\|_p,$$

where C is independent of Ω.

Proof. First, let $f \in C_c^\infty(\Omega)$ and let $\Omega_R = \Omega \cap C_R$ where, as before, C_R is a cylinder centered at the origin of height $2R$ with axis in the graph direction and with a base of diameter R. Then Ω_R is a bounded convex domain and for $R > 0$ sufficiently large such that $\text{supp} f \subseteq \Omega_R$, we can solve

$$(2.14) \qquad \begin{cases} \Delta u_R = f & \text{in } \Omega_R, \\ u_R = 0 & \text{on } \partial\Omega_R \end{cases}$$

and we get $\|u_R\|_{2,p} \leq C_R \|f\|_p$ using Theorem 2.1.2.

Now let $\Omega_1^R = \{x \in \mathbb{R}^n : Rx \in \Omega_R\}$, $f_1(x) = f(Rx)$ and $u_1(x) = u_R(Rx)$ for $x \in \Omega_1^R$. Then u_1 solves

$$(2.15) \qquad \begin{cases} \Delta u_1 = R^2 f_1 & \text{in } \Omega_1^R, \\ u_1 = 0 & \text{on } \partial\Omega_1^R \end{cases}$$

and by Theorem 2.1.2, we have that

(2.16) $\|u_1\|_{W^{2,p}(\Omega_1^R)} \leq C R^2 \|f_1\|_{L^p(\Omega_1^R)}$

and C is independent of Ω and R as the diameter of Ω_1^R is uniformly bounded
in R. Now,

$$\begin{aligned}
\|u_1\|_{L^p(\Omega^R)}^p &= R^{-n} \|u_R\|_{L^p(\Omega)}^p, \\
\|\partial_i u_1\|_{L^p(\Omega^R)}^p &= R^{-n+p} \|\partial_i u_R\|_{L^p(\Omega)}^p, \\
\|\partial_i \partial_j u_1\|_{L^p(\Omega^R)}^p &= R^{-n+2p} \|\partial_i \partial_j u_R\|_{L^p(\Omega)}^p.
\end{aligned}$$

For the solution of (2.14) this gives

$$\|u_R\|_p + R \|\nabla u_R\|_p + R^2 \|\nabla^2 u_R\|_p \leq C R^2 \|f\|_p.$$

This implies that the norm $\|\nabla^2 u_R\|_p$ is uniformly bounded and we obtain
$D_i D_j u_R \rightharpoonup u_{ij}$ in $L^p(\Omega)$ for some functions $u_{ij} \in L^p(\Omega)$, $1 \leq i, j \leq n$.

Locally we have the following result: Extend f trivially to $\mathbb{R}^n$ and define the
function $h = (h_j)_{j=1,\ldots,n}$, via its Fourier transform by

$$\mathcal{F} h_j(\xi) := \frac{i \xi_j}{|\xi|^2} \mathcal{F} f(\xi).$$

Then $\mathcal{F} f(\xi) = -i \sum_{j=1}^n \xi_j \mathcal{F} h_j(\xi)$, i.e. $f = \operatorname{div} h$. Since $f \in C_c^\infty(\mathbb{R}^n)$, we
have $\mathcal{F} f \in \mathcal{S}(\mathbb{R}^n)$. Therefore, h is in $L^2(\mathbb{R}^n)$ if $\xi \mapsto \xi_j |\xi|^{-2}$ is square inte-
grable near the origin. This is the case here, as $n > 2$. Now, recalling that
since $f \in C_c^\infty(\Omega)$ we have $u_R \in H^2(\Omega)$, we get

$$-\int_\Omega h \nabla u_R = \int_\Omega f u_R = -\int_\Omega |\nabla u_R|^2,$$

so

$$\|\nabla u_R\|_{L^2(\Omega)} \leq \|h\|_{L^2(\Omega)} =: C$$

and we find a subsequence of the $(\nabla u_R)_{R>0}$ converging weakly in $L^2(\Omega)$.
Then Poincaré's inequality on bounded subsets implies that the subsequence
in fact converges in $H^1_{\text{loc}}(\Omega)$, i.e. $u_R \rightharpoonup u$ in $H^1(\Omega_b)$ for any bounded $\Omega_b \subseteq \Omega$.
Then for $\varphi \in C_c^\infty(\Omega)$ and R sufficiently large such that $\operatorname{supp} \varphi \subseteq \Omega_R$, we get

$$\int_\Omega u_{ij} \varphi = \lim_{R \to \infty} \int_\Omega D_i D_j u_R \varphi$$

$$\begin{aligned}
&= \lim_{R \to \infty} \int_{\text{supp } \varphi} D_i D_j u_R \varphi \\
&= \lim_{R \to \infty} \int_{\text{supp } \varphi} u_R D_i D_j \varphi \\
&= \int_{\text{supp } \varphi} u D_i D_j \varphi \\
&= \int_{\Omega} u D_i D_j \varphi = \int_{\Omega} D_i D_j u \varphi,
\end{aligned}$$

so $D_i D_j u = u_{ij} \in L^p(\Omega)$. As for the gradient in the case of an arbitrary Lipschitz domain, using the weak convergence of the functions u_R, we get

$$\left\| \nabla^2 u \right\|_p \leq C \left\| f \right\|_p.$$

Furthermore, $0 = \widetilde{\text{tr}}\, u_R \rightharpoonup \widetilde{\text{tr}}\, u$ implies $\widetilde{\text{tr}}\, u = 0$, i.e. the trace of u is constant and we can choose the representative u such that $\text{tr}\, u = 0$.

For the general case, when $f \in L^p(\Omega)$, approximate f by $f_n \in C_c^\infty(\Omega)$. Then solve

$$\begin{cases}
\Delta u_n = f_n & \text{in } \Omega, \\
u_n = 0 & \text{on } \partial\Omega.
\end{cases}$$

We have

$$\left\| \nabla^2 u_n \right\|_p \leq C \left\| f_n \right\|_p \leq C.$$

Therefore, $D_i D_j u_n \rightharpoonup u_{ij}$ in $L^p(\Omega)$. We can now proceed as in the case of smooth f to obtain a solution u satisfying the boundary condition and the estimate on the second derivatives. $\qquad \square$

Chapter 3

Maximal regularity for the Laplacian on Lipschitz domains

In this chapter we will investigate solutions to the equations

$$
\begin{cases}
u' - \Delta u = f & \text{in } \Omega, \\
u = 0 & \text{on } \partial\Omega, \\
u(0) = 0 & \text{in } \Omega,
\end{cases}
\qquad \text{and} \qquad
\begin{cases}
u' - \Delta u = f & \text{in } \Omega, \\
\frac{\partial u}{\partial N} = 0 & \text{on } \partial\Omega, \\
u(0) = 0 & \text{in } \Omega.
\end{cases}
$$

For bounded smooth domains Ω, it is well-known that the Laplacian with domain $D(\Delta) = W^{2,p}(\Omega) \cap W_0^{1,p}(\Omega)$ generates an analytic C_0-semigroup of contractions on $L^p(\Omega)$, $1 < p < \infty$, (see e.g. [Paz83, Section 7.3]) and that it has the maximal regularity property (this follows e.g. from [DV87]). Similarly, in the bounded smooth case, the domain for the Neumann-Laplacian lies in $W^{2,p}(\Omega)$ (see e.g. [Gri85, Section 2.4]).

However, in Lipschitz domains this is no longer the case as we will see in the final section of this chapter where we use results on harmonic functions in the complement of slender cones to construct a bounded Lipschitz domain Ω_1 and an unbounded smooth domain Ω_2 such that the Laplacian with domains $D_1(\Delta) = \{u \in W_0^{1,p}(\Omega_i) : \Delta u \in L^p(\Omega_i)\}$ and $D_2(\Delta) = W^{2,p}(\Omega_i) \cap W_0^{1,p}(\Omega_i)$, $i = 1, 2$, isn't even a closed operator.

In the first section of this chapter, we will study the Laplacian with Dirichlet boundary conditions with the two different domains $D_1(\Delta)$ and $D_2(\Delta)$.

We will show that, under suitable assumptions on the Lipschitz domain Ω and the exponent p, these operators generate C_0-semigroups in $L^p(\Omega)$ which are positive, analytic and contractive. In particular, in the case of bounded convex domains Ω and $1 < p \le 2$, we are able to prove the same results as in the smooth case. Moreover, the generated semigroups coincide with the semigroups gained by the form method (see e.g. [Ouh04]). In the second section, we repeat these arguments and obtain similar results for the Neumann-Laplacian.

The results of the first two sections are then combined in the third section with results from Chapter 1 to obtain our main result: maximal regularity for the Laplacian in Lipschitz domains under the same assumptions on the domain Ω and the exponent p needed in the first two sections for generation of a semigroup.

As mentioned above, the final section then introduces a counterexample proving that extra conditions are necessary on Lipschitz domains to even guarantee $W^{1,p}$-solutions to the heat equation.

3.1 Generation of a C_0-semigroup by the Dirichlet-Laplacian

The aim of this section is to show that if we define the Dirichlet-Laplacian suitably on the Lebesgue spaces $L^p(\Omega)$, for certain domains Ω and a range of exponents p, the Dirichlet-Laplacian is the generator of a C_0-semigroup on $L^p(\Omega)$. We then determine various properties of the generated semigroup. To begin, we introduce the Dirichlet-Laplacian with two different domains of definition. The first of the operators to be introduced is the weak Dirichlet-Laplacian.

Definition 3.1.1. *We define the* weak Dirichlet-Laplacian $\Delta_{p,w}^D$ *on* $L^p(\Omega)$ *by*

$$D(\Delta_{p,w}^D) = \{u \in W_0^{1,p}(\Omega) : \Delta u \in L^p(\Omega)\},$$
$$\Delta_{p,w}^D u = \Delta u.$$

Here, $\Delta u \in L^p(\Omega)$ *is to be understood in the sense of distributions.*

In order to obtain results on higher regularity of the solution to the Cauchy problem, we introduce the strong Dirichlet-Laplacian.

Definition 3.1.2. *The* strong Dirichlet-Laplacian $\Delta_{p,s}^D$ *on* $L^p(\Omega)$ *is defined by*

$$D(\Delta_{p,s}^D) = W^{2,p}(\Omega) \cap W_0^{1,p}(\Omega),$$
$$\Delta_{p,s}^D f = \Delta f.$$

Remarks 3.1.3. Obviously, both of these operators are densely defined, as $C_c^\infty(\Omega)$ is dense in $L^p(\Omega)$ and is contained in the domain of the operators.

Our aim is to use the Lumer-Phillips Theorem to prove that the Dirichlet-Laplacian generates a C_0-semigroup. To apply the theorem, we need to show dissipativity of the operator (cf. Definition 1.3.3).

Lemma 3.1.4. *Let* $\Omega \subseteq \mathbb{R}^n$ *be a bounded Lipschitz domain and* $2 \leq p < \infty$. *Then* $\Delta_{p,w}^D$ *is dissipative.*

Proof. Let $u \in D(\Delta_{p,w}^D)$ and set

$$u^*(x) = \begin{cases} |u(x)|^{p-2}\overline{u}(x) & \text{if } u(x) \neq 0, \\ 0 & \text{otherwise.} \end{cases}$$

Then $u^* \in L^{p'}(\Omega)$ and $\widetilde{u}^* := \frac{u^*}{\|u^*\|_{p'}} \in dN(u)$, the subdifferential of u. We also have

$$\nabla u^* = |u|^{p-2}\nabla\overline{u} + \frac{p-2}{2}\overline{u}|u|^{p-4}\left(\overline{u}\nabla u + u\nabla\overline{u}\right)$$

on the set $\{x : u(x) \neq 0\}$. Using Hölder's inequality, it is easy to check that $\nabla u^* \in L^{p'}(\Omega)$. Moreover, by the discussion in Section 1.2.4, as $u^* \in W^{1,p'}(\Omega)$, tr u^* exists in $L^{p'}(\partial\Omega)$. If $u \in C(\overline{\Omega})$, then $u^* \in C(\overline{\Omega})$ and

$$(\text{tr } u)^p = (u|_{\partial\Omega})^p = u^p|_{\partial\Omega} = (u^*)^{p'}|_{\partial\Omega} = (u^*|_{\partial\Omega})^{p'} = (\text{tr } u^*)^{p'},$$

so tr $u^* = 0$. By density, we have tr $u^* = 0$ for all $u \in W_0^{1,p}(\Omega)$ and therefore, by Proposition 1.2.18, we have $u^* \in W_0^{1,p'}(\Omega)$. We can now integrate by parts using Corollary 1.2.24 to obtain

$$(3.1) \qquad \text{Re}\,\langle \Delta u, u^* \rangle = -\text{Re}\int_\Omega \nabla u \cdot \nabla u^* \, \chi_{\{u \neq 0\}}.$$

A calculation then yields

$$
\begin{aligned}
\operatorname{Re}\langle \Delta u, u^* \rangle &= -\operatorname{Re}\int_\Omega \nabla u \cdot \nabla u^* \, \chi_{\{u \neq 0\}} \\
&= -\operatorname{Re}\int_\Omega |u|^{p-2}|\nabla u|^2 \chi_{\{u \neq 0\}} \\
&\qquad -\operatorname{Re}\int_\Omega \frac{p-2}{2}\overline{u}|u|^{p-4}\left(\overline{u}(\nabla u)^2 + u|\nabla u|^2\right)\,\chi_{\{u \neq 0\}} \\
&= -\int_\Omega \left(|u|^{p-2}|\nabla u|^2 + (p-2)|u|^{p-4}(\operatorname{Re}\,\overline{u}\nabla u)^2\right)\,\chi_{\{u \neq 0\}} \\
&= -\int_\Omega |u|^{p-4}\left(\overline{u}\nabla u \cdot u\nabla\overline{u} + (p-2)(\operatorname{Re}\,\overline{u}\nabla u)^2\right)\,\chi_{\{u \neq 0\}} \\
&= -\int_\Omega |u|^{p-4}\left[(p-1)(\operatorname{Re}\,\overline{u}\nabla u)^2 + (\operatorname{Im}\,\overline{u}\nabla u)^2\right]\,\chi_{\{u \neq 0\}} \leq 0. \quad \square
\end{aligned}
$$

As a consequence, we have

Corollary 3.1.5. *Let $\Omega \subseteq \mathbb{R}^n$ be a bounded Lipschitz domain and $2 \leq p < \infty$. Then $\Delta_{p,s}^D$ is dissipative.*

Proof. We need to show that $\operatorname{Re}\langle \Delta u, u^* \rangle \leq 0$ for any $u \in D(\Delta_{p,s}^D)$. However, the strong Dirichlet-Laplacian is contained in the weak Dirichlet-Laplacian and so this holds by the previous lemma. $\square$

Note that for $p < 2$, the function u^* is not in $W_0^{1,p'}(\Omega)$, so the straightforward integration by parts is not possible. However, for the strong Dirichlet-Laplacian an approximation procedure yields the desired result:

Lemma 3.1.6. *Let Ω be a Lipschitz domain, $1 < p < \infty$ and $u^* = |u|^{p-2}\overline{u}$. Then for $u \in W^{2,p}(\Omega)$ we have*

$$
(3.2) \quad \int_\Omega \Delta u \, u^* = -(p-1)\int_\Omega |u|^{p-4}|\operatorname{Re}(\overline{u}\nabla u)|^2 \chi_{\{u \neq 0\}}
$$
$$
- \int_\Omega |u|^{p-4}|\operatorname{Im}(\overline{u}\nabla u)|^2 \chi_{\{u \neq 0\}}
$$
$$
- i(p-2)\int_\Omega |u|^{p-4}\operatorname{Re}(\overline{u}\nabla u)\operatorname{Im}(\overline{u}\nabla u)\chi_{\{u \neq 0\}}
$$
$$
+ \int_{\partial\Omega} \overline{u}|u|^{p-2}\frac{\partial u}{\partial N}.
$$

Proof. This is a special case of Theorem 3.1 from [MS05] with $\phi = 0$ and $A = \Delta$. Note that the assumptions on the boundary in [MS05] only require that $C^\infty(\overline{\Omega})$ is dense in $W^{2,p}(\Omega)$ and that traces are well defined. In particular the results hold for Lipschitz domains. $\square$

Corollary 3.1.7. *Let $\Omega \subseteq \mathbb{R}^n$ be a Lipschitz domain and $1 < p < \infty$. Then $\Delta_{p,s}^D$ is dissipative.*

Proof. Taking real parts in (3.2), we see that the third integral disappears. The boundary integral vanishes due to the boundary condition. The remaining terms are obviously negative. $\square$

We are now in the position to prove one of the main theorems of this section for the strong Dirichlet-Laplacian.

Theorem 3.1.8. *Let $\Omega \subseteq \mathbb{R}^n$, $n \geq 2$, be a bounded Lipschitz domain satisfying a uniform outer ball condition and $1 < p \leq 2$. Then $\Delta_{p,s}^D$ generates a C_0-semigroup of contractions on $L^p(\Omega)$.*

Proof. It remains to verify the range condition of Theorem 1.3.5, i.e. that

$$(\lambda - \Delta)D(\Delta_{p,s}^D) = L^p(\Omega) \text{ for some } \lambda > 0$$

is satisfied. However, from Theorem 2.1.2, we know that under our assumptions, $0 \in \rho(\Delta_{p,s}^D)$ and as the resolvent set is open, we have $\lambda \in \rho(\Delta_{p,s}^D)$ for some small $\lambda > 0$ which proves the theorem. $\square$

For the weak Dirichlet-Laplacian we obtain the following result:

Theorem 3.1.9. *Let $\Omega \subseteq \mathbb{R}^n$, $n \geq 3$, be a bounded Lipschitz domain. Then there exists $\varepsilon > 0$ depending only on the Lipschitz constant of Ω such that the operator $\Delta_{p,w}^D$ generates a C_0-semigroup of contractions on $L^p(\Omega)$ for $(3+\varepsilon)' < p < 3+\varepsilon$, where $(3+\varepsilon)'$ denotes the conjugate exponent to $3+\varepsilon$, . If $n = 2$, the same statement holds for all $(4+\varepsilon)' < p < 4+\varepsilon$.*

If Ω is a bounded Lipschitz domain in $\mathbb{R}^n$, $n \geq 2$, satisfying a uniform outer ball condition, the statement is true for all $1 < p < \infty$.

Proof. We start with the case $n \geq 3$. Once again, we verify that the range condition of Theorem 1.3.5 is satisfied, i.e.

$$(\lambda - \Delta)D(\Delta_{p,w}^{D}) = L^p(\Omega) \text{ for some } \lambda > 0.$$

For $u \in D(\Delta_{p,w}^{D})$, we have $\Delta u \in L^p(\Omega)$, so $\Delta u \in L_{\alpha-2}^{p}(\Omega)$ for any $\alpha \leq 2$. Whenever $1 < p < 3 + \varepsilon$, we can find $\alpha \in [1, 2]$ and a unique $v \in L_{\alpha}^{p}(\Omega)$ such that

$$\begin{cases} \Delta v = \Delta u & \text{in } \Omega, \\ \quad v = 0 & \text{on } \partial\Omega, \end{cases}$$

and $\|v\|_{\alpha,p} \leq C \|\Delta u\|_{\alpha-2,p}$ by Theorem 2.1.1. Then $u - v$ is harmonic and by the Maximum Principle we have $u = v$. Thus

$$\|u\|_{1,p} \leq \|u\|_{\alpha,p} \leq C \|\Delta u\|_{\alpha-2,p} \leq C \|\Delta u\|_{p}.$$

Therefore, $\Delta_{p,w}^{D} : D(\Delta_{p,w}^{D}) \to L^p(\Omega)$ is an injective mapping for our range of exponents p. A similar argument shows that it is also surjective. Thus under our assumptions we have $0 \in \rho(\Delta_{p,w}^{D})$ for any $1 < p < 3 + \varepsilon$. Moreover, by Lemma 3.1.4, $\Delta_{p,w}^{D}$ is dissipative for $p \geq 2$. This proves the theorem for $2 \leq p < 3 + \varepsilon$.

For the case when $(3+\varepsilon)' < p < 2$ we consider the dual operator. Since $\Delta_{p',w}^{D}$ is m-dissipative (i.e. is dissipative and satisfies the range condition), its dual operator $\Delta_{p',w}^{D}{}'$ is m-dissipative in $L^p(\Omega)$ (cf. [CH87, Proposition 3.10][1]). We now claim that $\Delta_{p,w}^{D} \subseteq \Delta_{p',w}^{D}{}'$, i.e. $D(\Delta_{p,w}^{D}) \subseteq D(\Delta_{p',w}^{D}{}')$ and both operators coincide on $D(\Delta_{p,w}^{D})$. To see this, let $v \in D(\Delta_{p,w}^{D})$, $u \in D(\Delta_{p',w}^{D})$, $(v_n) \subseteq C_c^{\infty}(\Omega)$ and $(u_n) \subseteq C_c^{\infty}(\Omega)$ such that $v_n \to v$ in $W^{1,p}(\Omega)$ and $u_n \to u$ in $W^{1,p'}(\Omega)$. Then

$$\begin{aligned} \langle \Delta u, v \rangle &= \lim_{n\to\infty} \langle \Delta u, v_n \rangle = -\lim_{n\to\infty} \langle \nabla u, \nabla v_n \rangle \\ &= -\langle \nabla u, \nabla v \rangle = -\lim_{n\to\infty} \langle \nabla u_n, \nabla v \rangle \\ &= \lim_{n\to\infty} \langle u_n, \Delta v \rangle = \langle u, \Delta v \rangle, \end{aligned}$$

---

[1]Note that the result is only stated for real Banach spaces in [CH87], but is also valid in our case using the same proof.

so $v \in D(\Delta_{p',w}^{D}{}')$ and $\Delta_{p',w}^{D}{}'v = \Delta v$ as claimed. Therefore $\Delta_{p,w}^{D}$ is contained in a dissipative operator and hence is itself dissipative for $(3 + \varepsilon)' < p \leq 2$. Moreover, as we have seen above, for these p the range condition is satified. Using the Lumer-Phillips Theorem, this completes the proof for $n \geq 3$.

For $n = 2$, we merely replace Theorem 2.1.1 by [JK95, Theorem 1.3], while in the case of domains satisfying a uniform outer ball condition we use Theorem 2.1.2 and argue in the same way obtaining the larger range of exponents p.

$\square$

Corollary 3.1.10. *If* $\Omega \subseteq \mathbb{R}^n$, $n \geq 2$, *is a bounded Lipschitz domain safisfying a uniform outer ball condition and* $1 < p \leq 2$, *then we have* $\Delta_{p,w}^{D} = \Delta_{p,s}^{D}$.

Proof. Obviously, $\Delta_{p,s}^{D} \subseteq \Delta_{p,w}^{D}$. Furthermore, $0 \in \rho(\Delta_{p,s}^{D}) \cap \rho(\Delta_{p,w}^{D})$.

We claim that if A and B are operators on a Banach space X such that $A \subseteq B$ and $\rho(A) \cap \rho(B) \neq \emptyset$, then $A = B$.

We have that $(\lambda - A) : D(A) \to X$ is bijective and that $(\lambda - B) : D(B) \to X$ is bijective. Since $(\lambda - B)|_{D(A)} = (\lambda - A)$ is already surjective, we must have $D(B) = D(A)$. $\square$

Remarks 3.1.11. 1. In particular, this proves that the weak Laplacian is dissipative in $L^p(\Omega)$ in bounded Lipschitz domains satisfying a uniform outer ball condition for all $1 < p < 2$.

2. To make the statements more concise, in the following we will often only refer to the semigroup generated by the weak Dirichlet-Laplacian, recalling that whenever the strong Dirichlet-Laplacian is a generator, it coincides with the weak Dirichlet-Laplacian.

We can show that the semigroups generated on $L^p(\Omega)$ are consistent.

Proposition 3.1.12. *Let* Ω *be a bounded Lipschitz domain in* $\mathbb{R}^n$, *where either*

- $n \geq 3$ *and* $(3 + \varepsilon)' < p, q < 3 + \varepsilon$, *where* $\varepsilon > 0$ *depends only on the Lipschitz constant of* Ω,

- $n = 2$ and $(4 + \varepsilon)' < p, q < 4 + \varepsilon$, where $\varepsilon > 0$ depends only on the Lipschitz constant of Ω or

- $n \geq 2$ and suppose additionally that Ω satisfies a uniform outer ball condition and $1 < p, q < \infty$.

Then the semigroups T_p generated by $\Delta_{p,w}^D$ and T_q generated by $\Delta_{q,w}^D$ are consistent, i.e. if $f \in L^p(\Omega) \cap L^q(\Omega)$ then

$$T_p(t)f = T_q(t)f \text{ for all } t \geq 0.$$

Proof. W.l.o.g. assume $p < q$. Then, as Ω is bounded, $D(\Delta_{q,w}^D) \subseteq D(\Delta_{p,w}^D)$. Let $f \in D(\Delta_{q,w}^D)$. Then $T_q(t)f$ is the unique classical solution to

$$(3.3) \qquad\qquad u' - \Delta u = 0, \ u(0) = f$$

with $T_q(t)f \in D(\Delta_{q,w}^D)$ for $t \geq 0$. But then $T_q(t)f \in D(\Delta_{p,w}^D)$, so it must agree with $T_p(t)f$, the unique classical solution to (3.3) for $f \in D(\Delta_{p,w}^D)$. Since $D(\Delta_{q,w}^D)$ is dense in $L^q(\Omega)$, we get $T_p(t)f = T_q(t)f$ for all $f \in L^q(\Omega)$. $\square$

This now gives us some further interesting results.

Corollary 3.1.13. *Let Ω, n and p be as in Proposition 3.1.12. Then the semigroup generated by $\Delta_{p,w}^D$ satisfies a Gaussian estimate, i.e. there exist constants $a \geq 0$, $M, b > 0$ such that*

$$|T(t)f| \leq Me^{at}G(bt)|f| \text{ for } t \geq 0,$$

where

$$G(t)f \ = \ k_t * f$$

is the Gaussian semigroup and

$$k_t(x) \ = \ (4\pi t)^{-\frac{n}{2}} e^{-\frac{|x|^2}{4t}}$$

is the Gaussian kernel.

Moreover, the semigroup generated by $\Delta_{p,w}^D$ is analytic.

Proof. On $L^2(\Omega)$, $\Delta_{2,w}^D$ is identical to the Dirichlet-Laplacian defined via the form in [AB99, Section 1]. Since the semigroups generated on $L^p(\Omega)$ are consistent, both for the construction of the generator in [AB99] and for the semigroups generated by $\Delta_{p,w}^D$, the semigroups must coincide for all cases. Then by [AB99, Section 1] the semigroup satisfies a Gaussian estimate and therefore, by [Ouh95], the semigroup on $L^p(\Omega)$ is analytic. $\qquad\square$

Corollary 3.1.14. *Let Ω, n and p be as in Proposition 3.1.12. Then the spectrum of $\Delta_{p,w}^D$ is independent of p, i.e. for all p such that $\Delta_{p,w}^D$ generates a C_0-semigroup in $L^p(\Omega)$, we have $\sigma(\Delta_{p,w}^D) = \sigma(\Delta_{2,w}^D)$.*

Proof. The resolvent of $\Delta_{p,w}^D$ maps $L^p(\Omega)$ to $W^{1,p}(\Omega)$ which is compactly embedded in $L^p(\Omega)$, so $\Delta_{p,w}^D$ has compact resolvent. Furthermore, the semigroups generated by $\Delta_{p,w}^D$ and $\Delta_{2,w}^D$ are consistent by Proposition 3.1.12. Therefore, by [Are94, Proposition 2.6] we obtain that $\sigma(\Delta_{p,w}^D) = \sigma(\Delta_{2,w}^D)$. $\qquad\square$

Our next aim in this section is to show that the generated C_0-semigroup is positive. To prove this, we will use the following proposition (cf. [CH87, Corollary 7.15]):

Proposition 3.1.15. *Let X be a Banach lattice and $A : D(A) \to X$ a linear operator. Then the following are equivalent:*

  a) A generates a positive contraction semigroup,

  b) 1. A is densely defined,

      2. for some $\lambda > 0$, the operator $\lambda - A$ is surjective,

      3. A is dispersive, i.e. for all $x \in D(A)$, there exists $0 \le \Phi \in X'$ with $\|\Phi\|_{X'} \le 1$, $\langle x, \Phi \rangle = \|x^+\|_X$ and $\langle Ax, \Phi \rangle \le 0$, where x^+ denotes the positive part of x.

Lemma 3.1.16. *Let Ω, n and p be as in Proposition 3.1.12. Then the semigroup generated by $\Delta_{p,w}^D$ is positive in $L^p(\Omega)$.*

Proof. With the previous proposition, we can prove the lemma when dealing only with real-valued functions as, in this case, $L^p(\Omega)$ is a Banach lattice. As we know that $\Delta_{p,w}^D$ generates a semigroup on $L^p(\Omega)$, the first two conditions in b) are satisfied.

To check the third, we first consider the case $p \geq 2$. Assume $\|u^+\|_p \neq 0$, otherwise simply choose $\Phi = 0$. Let $\Phi = \frac{(u^+)^{p-1}}{\|u^+\|_p^{p/p'}}$ and integrate by parts as in the proof of Lemma 3.1.4 using that $D_i u^+ = (D_i u)\,\chi_{\{u>0\}}$ (see e.g. [GT77, Lemma 7.6]) and that tr $u^+ = 0$. To see the latter, write $2u^+ = u + |u|$, so it suffices to show tr $|u| = 0$. Now approximate u in $W^{1,p}(\Omega)$ by C_c^∞-functions u_n and set $v_n := \sqrt{u_n^2 + 1/n^2} - 1/n$. Then $v_n \in C_c^\infty(\Omega)$ and it can be shown that they converge to $|u|$ in $W^{1,p}(\Omega)$. This gives positivity of the semigroup whenever $p \geq 2$.

When dealing with complex-valued functions, the positivity of the semigroup on $L^p(\Omega)$, $p \geq 2$, obviously follows from the real-valued case and the fact that the operator has real-valued coefficients.

Now let $p < 2$. Recall that a semigroup is positive iff the resolvent is positive for sufficiently large $\lambda > 0$ (see e.g. [CH87, Proposition 7.1]). Let $f \in L^p(\Omega)$, $f \geq 0$. Then there exist $f_n \in L^2(\Omega)$, $f_n \geq 0$ such that $f_n \to f$ in $L^p(\Omega)$. If $\Delta_{p,w}^D$ generates a C_0-semigroup on $L^p(\Omega)$, then from the resolvent estimate we obtain that $R(\lambda, \Delta_{p,w}^D)f_n \to R(\lambda, \Delta_{p,w}^D)f$ in $L^p(\Omega)$. However, positivity of the semigroup on $L^2(\Omega)$ and consistency of the semigroups (Proposition 3.1.12) imply $R(\lambda, \Delta_{p,w}^D)f_n \geq 0$ and therefore $R(\lambda, \Delta_{p,w}^D)f \geq 0$ almost everywhere. This proves positivity also for the case $p < 2$. $\square$

We finish this section with a result on the growth bound of the generated semigroup.

Corollary 3.1.17. *Let Ω, n and p be as in Proposition 3.1.12. The semigroup generated by $\Delta_{p,w}^D$ is of negative type in $L^p(\Omega)$, i.e. the growth bound of the semigroup satisfies $\omega(T) < 0$, moreover, it is independent of p.*

Proof.   A result due to Weis (cf. [Wei95], [Wei98] or [ABHN01, Theorem 5.3.6] for different versions of the proof), implies that for generators A of positive semigroups T on $L^p(\Omega)$, we have $s(A) = \omega(T)$. By Corollary 3.1.14, we already know that $\sigma(\Delta_{p,w}^D) = \sigma(\Delta_{2,w}^D)$, in particular equality holds for the spectral bound. It therefore remains to examine the case $p = 2$.

Using Poincaré's inequality (Theorem 1.2.15), for $u \in D(\Delta_{2,w}^D)$ we have

$$\frac{\langle \Delta u, u \rangle}{\|u\|_2^2} = -\frac{\|\nabla u\|_2^2}{\|u\|_2^2} \leq -C < 0.$$

Then by [Paz83, Theorem 1.3.9], we get that the numerical range and therefore the spectrum of $\Delta_{2,w}^D$ lie in the half-plane $\{z \in \mathbb{C} : \operatorname{Re} z \leq -C\}$, in particular $s(\Delta_{2,w}^D) \leq -C$. □

3.2 Generation of a C_0-semigroup by the Neumann-Laplacian

For Neumann boundary conditions, we also introduce the weak and the strong Neumann-Laplacian. See Definition 1.2.25 for the precise meaning of the boundary conditions.

Definition 3.2.1. *We define the* weak Neumann-Laplacian $\Delta_{p,w}^N$ *on* $L^p(\Omega)$ *by*

$$D(\Delta_{p,w}^N) = \left\{ u \in W^{1,p}(\Omega) : \frac{\partial u}{\partial N} = 0 \text{ on } \partial\Omega, \ \Delta u \in L^p(\Omega) \right\},$$
$$\Delta_{p,w}^N u = \Delta u.$$

As usual, $\Delta u \in L^p(\Omega)$ *is to be understood in the sense of distributions.*

Definition 3.2.2. *The* strong Neumann-Laplacian $\Delta_{p,s}^N$ *is defined on* $L^p(\Omega)$ *by*

$$D(\Delta_{p,s}^N) = \left\{ f \in W^{2,p}(\Omega) : \frac{\partial f}{\partial N} = 0 \text{ on } \partial\Omega \right\},$$
$$\Delta_{p,s}^N f = \Delta f.$$

Remarks 3.2.3. As in the case of Dirichlet boundary conditions, both of these operators are obviously densely defined, as $C_c^\infty(\Omega)$ is dense in $L^p(\Omega)$ and is contained in the domain of the operators.

We will now proceed in a very similar way to the previous section to prove generator results for the Neumann-Laplacian. As solutions to the Neumann problem in bounded domains are only unique up to a constant, we introduce the space

$$L_0^p(\Omega) := \left\{ f \in L^p(\Omega) : \int_\Omega f = 0 \right\}$$

of L^p-functions with mean zero. Note that for bounded domains Ω and $f \in L^p(\Omega)$ the expression $\int_\Omega f$ is well-defined. Also, $L_0^p(\Omega)$ is a closed subspace of $L^p(\Omega)$ and therefore also a Banach space.

Once again, our aim is to use the Lumer-Phillips Theorem to prove generation of a C_0-semigroup. Therefore, we need to show dissipativity of the operator.

Lemma 3.2.4. *Let $\Omega \subseteq \mathbb{R}^n$ be a Lipschitz domain. Then the weak Neumann-Laplacian $\Delta_{p,w}^N$ is dissipative in $L^p(\Omega)$ for $2 \leq p < \infty$ and the strong Neumann-Laplacian $\Delta_{p,s}^N$ is for $1 < p < \infty$.*

Proof. First consider the case $2 \leq p < \infty$. Define u^* as in the proof of Lemma 3.1.4. Since $u^* \in W^{1,p'}(\Omega)$, by definition of the boundary condition, we have

$$\text{Re} \langle \Delta u, u^* \rangle = -\text{Re} \int_\Omega \nabla u \cdot \nabla u^*,$$

Then the same calculation as in the proof of Lemma 3.1.4 yields the result. In the case of the strong Laplacian and $1 < p < 2$, we take the real part in formula (3.2). $\qquad \square$

We can now prove generation of a C_0-semigroup for the strong Neumann-Laplacian.

Theorem 3.2.5. *Let $\Omega \subseteq \mathbb{R}^n$, $n \geq 3$, be a bounded convex domain and $1 < p \leq 2$. Then $\Delta_{p,s}^N$ generates a C_0-semigroup of contractions on $L^p(\Omega)$.*

Proof. It remains to verify the range condition of Theorem 1.3.5, i.e.

$$(\lambda - \Delta)D(\Delta_{p,s}^N) = L^p(\Omega) \text{ for some } \lambda > 0.$$

From Theorem 2.1.8, we know that if we restrict the operator $\Delta_{p,s}^N$ to the space $L_0^p(\Omega)$ (again denoting the restriction by $\Delta_{p,s}^N$), we have $0 \in \rho(\Delta_{p,s}^N)$ and as the resolvent set is open, we have $\lambda \in \rho(\Delta_{p,s}^N)$ for some small $\lambda > 0$, i.e. for each $f \in L_0^p(\Omega)$, we can find $u \in W^{2,p}(\Omega) \cap L_0^p(\Omega)$ satisfying the Neumann boundary conditions such that $(\lambda - \Delta)u = f$. Given any $f \in L^p(\Omega)$, we can then write $f = (f - \bar{f}) + \bar{f}$ where $\bar{f} := |\Omega|^{-1} \int_\Omega f$. We can then find $u_0 \in W^{2,p}(\Omega) \cap L_0^p(\Omega)$ satisfying the Neumann boundary conditions such that $(\lambda - \Delta)u_0 = f - \bar{f}$. Then $u := u_0 + \lambda^{-1}\bar{f} \in D(\Delta_{p,s}^N)$ and $(\lambda - \Delta)u = f$.
□

Remarks 3.2.6. Note that for the solution $u = u_0 + \lambda^{-1}\bar{f} \in D(\Delta_{p,s}^N)$ in the proof above, we have that u is a solution in $D(\Delta_{p,s}^N)$ to $\Delta u = \lambda u - f =: g$ with $g \in L_0^p(\Omega)$. Therefore, by Theorem 2.1.6, u is unique up to a constant. However, because of the boundary condition, we have $\int_\Omega \Delta u = 0$. Hence, for $\lambda > 0$, we must have $\bar{u} := |\Omega|^{-1} \int_\Omega u = \lambda^{-1}\bar{f}$ which fixes the constant uniquely and the solution u given in the proof above is the only solution to $(\lambda - \Delta)u = f$, i.e. the operator is also injective and so $(0, \infty) \subseteq \rho(\Delta_{p,s}^N)$.

In the weak case, we obtain the following result:

Theorem 3.2.7. *Let $\Omega \subseteq \mathbb{R}^n$, $n \geq 3$, be a bounded Lipschitz domain. Then there exists $\varepsilon > 0$ depending only on the Lipschitz constant of Ω such that the operator $\Delta_{p,w}^N$ generates a C_0-semigroup of contractions on $L^p(\Omega)$ for $(3 + \varepsilon)' < p < 3 + \varepsilon$, where $(3 + \varepsilon)'$ denotes the conjugate exponent to $3 + \varepsilon$.*

Proof. Once again, we verify that the range condition of Theorem 1.3.5 is satisfied, i.e.

$$(\lambda - \Delta)D(\Delta_{p,w}^N) = L^p(\Omega) \text{ for some } \lambda > 0.$$

Proceeding as in the proof of Theorem 3.1.9 and using Theorem 2.1.6 we know that under our assumptions, for any $1 < p < 3 + \varepsilon$, the restriction of the operator $\Delta_{p,w}^N$ is invertible in $L_0^p(\Omega)$. As in the proof of the previous theorem we can then verify the range condition. Moreover, by Lemma 3.2.4, $\Delta_{p,w}^N$ is dissipative for $p \geq 2$. This proves the theorem for $2 \leq p < 3 + \varepsilon$.

For the case when $(3 + \varepsilon)' < p < 2$ we use that the dual operator $\Delta_{p',w}^N{}'$ is m-dissipative in $L^p(\Omega)$ and that $\Delta_{p,w}^N \subseteq \Delta_{p',w}^N{}'$. The latter follows directly from the definition of the boundary conditions, as for $v \in D(\Delta_{p,w}^N)$ and $u \in D(\Delta_{p',w}^N)$, we have

$$\langle \Delta u, v \rangle = -\langle \nabla u, \nabla v \rangle = \langle u, \Delta v \rangle.$$

Therefore $\Delta_{p,w}^N$ is contained in a dissipative operator and hence is itself dissipative for $(3 + \varepsilon)' < p \leq 2$. Moreover, as we have seen above, for these p the range condition is satified. Using the Lumer-Phillips Theorem this completes the proof. $\qquad\square$

Corollary 3.2.8. *If $\Omega \subseteq \mathbb{R}^n$, $n \geq 3$, is a bounded convex domain and $1 < p \leq 2$, we have $\Delta_{p,w}^N = \Delta_{p,s}^N$.*

Proof.   We note that, using the same argument as in Remarks 3.2.6 for the weak Neumann-Laplacian, we have $\lambda \in \rho(\Delta_{p,s}^N) \cap \rho(\Delta_{p,w}^N)$ for some small $\lambda > 0$. The proof is then analogous to that of Corollary 3.1.10 for the Dirichlet-Laplacian. $\qquad\square$

Remarks 3.2.9. 1. The weak Laplacian is dissipative in $L^p(\Omega)$ in bounded convex domains, not just for $2 \leq p < \infty$, but also for all $1 < p < 2$.

2. As before, to make the statements more concise, in the following we will often only refer to the semigroup generated by the weak Neumann-Laplacian, recalling that whenever the strong Neumann-Laplacian is a generator, it coincides with the weak Neumann-Laplacian.

We can show that the semigroups generated on $L^p(\Omega)$ are consistent.

Proposition 3.2.10. *Let Ω be a bounded Lipschitz domain in $\mathbb{R}^n$, $n \geq 3$, where either*

- $(3 + \varepsilon)' < p, q < 3 + \varepsilon$ *or*

- Ω *is convex and* $1 < p, q < 3 + \varepsilon$,

for some $\varepsilon > 0$. Then the semigroups T_p generated by $\Delta_{p,w}^N$ and T_q generated by $\Delta_{q,w}^N$ are consistent, i.e. if $f \in L^p(\Omega) \cap L^q(\Omega)$ then

$$T_p(t)f = T_q(t)f \text{ for all } t \geq 0.$$

Proof. See the proof of Proposition 3.1.12. □

Again, the semigroups generated by $\Delta_{p,w}^N$ agree with those obtained by the form method giving us further interesting results.

Corollary 3.2.11. *Let Ω and p be as in Proposition 3.2.10. Then the semigroup generated by $\Delta_{p,w}^D$ satisfies a Gaussian estimate and is analytic.*

Proof. On $L^2(\Omega)$, $\Delta_{2,w}^N$ is identical to the form Neumann-Laplacian Δ^N with domain

$$D(\Delta^N) \;=\; \left\{ u \in H^1(\Omega) : \exists v \in L^2(\Omega). \; \forall \varphi \in H^1(\Omega). \; \int_\Omega \nabla v \overline{\nabla \varphi} = -\int_\Omega v \overline{\varphi} \right\}$$

and $\Delta^N u = v$. By [Ouh04, Theorem 6.10], the generated semigroup satisfies a Gaussian estimate. Using this estimate, it is possible to construct a consistent family of semigroups on all $L^p(\Omega)$ (see [Ouh95]). Because of consistency, the semigroups agree with the ones we obtained in Theorem 3.2.7 and hence the Gaussian estimate holds for the whole range of p for which $\Delta_{p,w}^N$ is a generator. Then, by [Ouh95], the semigroup on $L^p(\Omega)$ is analytic. □

As in the case of Dirichlet boundary conditions, we have the following corollary on p-independence of the spectrum:

Corollary 3.2.12. *Let Ω and p be as in Proposition 3.2.10. Then the spectrum of $\Delta_{p,w}^N$ is independent of p, i.e. for all p such that $\Delta_{p,w}^N$ generates a C_0-semigroup in $L^p(\Omega)$, we have $\sigma(\Delta_{p,w}^N) = \sigma(\Delta_{2,w}^N)$.*

Finally, we show positivity of the generated C_0-semigroup.

Lemma 3.2.13. *Let Ω and p be as in Proposition 3.2.10. Then the semigroup generated by $\Delta_{p,w}^N$ is positive in $L^p(\Omega)$.*

Proof. By Proposition 3.1.15, in the real-valued case it is sufficient to check dispersiveness. For $p \geq 2$ and $\Phi = \frac{(u^+)^{p-1}}{\|u^+\|_p^{p/p'}}$, we have $\Phi \in W^{1,p'}(\Omega)$ and

$$\langle \Delta u, \Phi \rangle = -\langle \nabla u, \nabla \Phi \rangle$$

for all $u \in D(\Delta_{p,w}^N)$ and a calculation as in the proof of Lemma 3.1.4 yields the desired result.

The complex-valued case and the case $p < 2$ follow in the same way as for Dirichlet boundary conditions in Lemma 3.1.16.   □

3.3 Maximal regularity for the Laplacian in bounded Lipschitz domains

In the previous sections we have proven generator results for the Dirichlet- and Neumann-Laplacian and gathered various properties of the semigroups and their generators. We can now exploit these results to show the desired maximal regularity property for the Laplacian. We start with the case of Dirichlet boundary conditions.

3.3.1   Maximal regularity for the Dirichlet-Laplacian

Theorem 3.3.1. *Let Ω be a bounded Lipschitz domain in $\mathbb{R}^n$, where either*

- $n \geq 3$ *and* $(3 + \varepsilon)' < p < 3 + \varepsilon$, *where* $\varepsilon > 0$ *depends only on the Lipschitz constant of Ω,*

- $n = 2$ *and* $(4 + \varepsilon)' < p < 4 + \varepsilon$, *where* $\varepsilon > 0$ *depends only on the Lipschitz constant of Ω, or*

- $n \geq 2$ *and suppose additionally that Ω satisfies a uniform outer ball condition and $1 < p < \infty$.*

Then the weak Dirichlet-Laplacian as defined in Definition 3.1.1 has the maximal regularity property, i.e. for $1 < q < \infty$ and for every $f \in L^q(\mathbb{R}_+, L^p(\Omega))$ there exists a unique solution to

(3.4)
$$\begin{cases} u'(t) - \Delta u(t) = f(t) & \text{for } t \in \mathbb{R}_+, \\ u(t, x) = 0 & \text{on } \mathbb{R}_+ \times \partial\Omega, \\ u(0) = 0. \end{cases}$$

The solution u lies in $L^q(\mathbb{R}_+, D(\Delta^D_{p,w})) \cap W^{1,q}(\mathbb{R}_+, L^p(\Omega))$ and satisfies the estimate

$$\|u\|_{L^q(\mathbb{R}_+, L^p(\Omega))} + \|u'\|_{L^q(\mathbb{R}_+, L^p(\Omega))} + \|\Delta u\|_{L^q(\mathbb{R}_+, L^p(\Omega))} \leq C \|f\|_{L^q(\mathbb{R}_+, L^p(\Omega))}.$$

Proof. We have shown that the semigroup generated by $\Delta_{p,w}^D$ is contractive (Theorems 3.1.8 and 3.1.9), analytic (Corollary 3.1.13) and positive (Lemma 3.1.16) on $L^p(\Omega)$. Maximal regularity now follows from Theorem 1.3.23. Moreover, as the generated semigroup is of negative type (Corollary 3.1.17), we get $u \in L^q(\mathbb{R}_+, L^p(\Omega))$ (cf. [Dor93]). $\qquad\square$

Of course, whenever the weak and the strong Laplacian coincide, this also yields maximal regularity for the strong Laplacian. Because of the importance of the result, in particular the better estimate (3.5), we state it here separately.

Theorem 3.3.2. *For $1 < p \leq 2$, and for all bounded Lipschitz domains $\Omega \subseteq \mathbb{R}^n$, with $n \geq 2$, and Ω satisfying a uniform outer ball condition, the strong Dirichlet-Laplacian as defined in Definition 3.1.2 has the maximal regularity property in $L^p(\Omega)$, i.e. for $1 < q < \infty$ and for every $f \in L^q(\mathbb{R}_+, L^p(\Omega))$ there exists a unique solution to*

$$\begin{cases} u'(t) - \Delta u(t) = f(t) & \text{for } t \in \mathbb{R}_+, \\ \quad u(t,x) = 0 & \text{on } \mathbb{R}_+ \times \partial\Omega, \\ \quad u(0) = 0. \end{cases}$$

The solution u lies in $L^q(\mathbb{R}_+, W^{2,p}(\Omega) \cap W_0^{1,p}(\Omega)) \cap W^{1,q}(\mathbb{R}_+, L^p(\Omega))$ and satisfies the estimate

$$(3.5) \qquad \|u\|_{L^q(\mathbb{R}_+, W^{2,p}(\Omega))} + \|u'\|_{L^q(\mathbb{R}_+, L^p(\Omega))} \leq C \|f\|_{L^q(\mathbb{R}_+, L^p(\Omega))} .$$

3.3.2 Maximal regularity for the Neumann-Laplacian

Similar results can now also be obtained for the Neumann-Laplacian. The main difference being that 0 is in the spectrum of the operator and so the solution itself will not lie in $L^q(\mathbb{R}_+, D(\Delta_p^N))$ (cf. [Dor93, Theorem 2.1]).

Theorem 3.3.3. *Let Ω be a bounded Lipschitz domain in $\mathbb{R}^n$, $n \geq 3$, where either*

- $(3 + \varepsilon)' < p < 3 + \varepsilon$ *or*

- Ω *is convex and $1 < p < 3 + \varepsilon$,*

for some $\varepsilon > 0$. Then the weak Neumann-Laplacian as defined in Definition 3.2.1 has the maximal regularity property, i.e. for $1 < q < \infty$ and for every $f \in L^q(\mathbb{R}_+, L^p(\Omega))$ there exists a unique solution to

(3.6)
$$\begin{cases} u'(t) - \Delta u(t) = f(t) & \text{for } t \in \mathbb{R}_+, \\ \frac{\partial u}{\partial N}(t, x) = 0 & \text{on } \mathbb{R}_+ \times \partial\Omega, \\ u(0) = 0. \end{cases}$$

The solution u satisfies the estimate

$$\|u'\|_{L^q(\mathbb{R}_+, L^p(\Omega))} + \|\Delta u\|_{L^q(\mathbb{R}_+, L^p(\Omega))} \leq C \|f\|_{L^q(\mathbb{R}_+, L^p(\Omega))}.$$

Proof.   Maximal regularity follows from contractivity (Theorems 3.2.5 and 3.2.7), analyticity (Corollary 3.2.11) and positivity (Lemma 3.2.13) of the semigroup generated by $\Delta_{p,w}^N$ and Theorem 1.3.23.   □

In convex domains we can also control the second derivatives.

Theorem 3.3.4. *For $1 < p \leq 2$, and for all bounded convex domains $\Omega \subset \mathbb{R}^n$ with $n \geq 3$, the strong Neumann-Laplacian as defined in Definition 3.2.2 has the maximal regularity property in $L^p(\Omega)$, i.e. for $1 < q < \infty$ and for every $f \in L^q(\mathbb{R}_+, L^p(\Omega))$ there exists a unique solution to (3.6). The solution u satisfies the estimate*

$$\|u'\|_{L^q(\mathbb{R}_+, L^p(\Omega))} + \left\|\nabla^2 u\right\|_{L^q(\mathbb{R}_+, L^p(\Omega))} \leq C \|f\|_{L^q(\mathbb{R}_+, L^p(\Omega))}.$$

3.4   Negative results

In this part, we will construct an example that proves that for the weak Dirichlet-Laplacian the upper bound on the exponent p given in Theorem 3.3.1 is optimal in bounded Lipschitz domains in the sense that for any $p > 3$, we can find a bounded Lipschitz domain Ω such that the operator $\Delta_{p,w}^D$ is not closed in $L^p(\Omega)$. Note that it follows easily from Theorem 2.1.1 that, given a Lipschitz domain Ω, there exists $\varepsilon > 0$ such that $\Delta_{p,w}^D$ is a closed operator in $L^p(\Omega)$ for $1 < p < 3 + \varepsilon$. In order to construct the example, we first make some observations on harmonic functions in the complement of a slender cone. Of particular interest will be the behaviour near the tip of the cone. The results that we will need are stated in the next theorem.

Theorem 3.4.1. *Let* $\Gamma_\varepsilon = \{x \in \mathbb{R}^3 : (x_1^2 + x_2^2)^{\frac{1}{2}} \leq -\varepsilon x_3\}$, $\varepsilon > 0$, *in* $\mathbb{R}^3$. *Consider the Dirichlet problem*

(3.7)
$$\begin{cases} \Delta u = 0 & \text{in } \Gamma_\varepsilon^c, \\ u = 0 & \text{on } \partial \Gamma_\varepsilon. \end{cases}$$

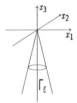

Then there is a solution to (3.7) of the form

$$u(r, \omega) = r^\lambda \phi(\omega),$$

where $r > 0$, $\lambda > 0$, $\omega \in \mathbb{S}^2$ *and* $\phi \neq 0$ *is a smooth function on the sphere. Furthermore,* $\lambda \to 0$ *as* $\varepsilon \to 0$ *and* u *is positive in* Γ_ε^c.

Proof. For these results see [KMR01, Sections 2.2 and 2.5.1]. □

Now consider the function v defined by $v(r, \omega) := u(r, \omega)\Theta(r)$ for $r > 0$ and $\omega \in \mathbb{S}^2$ where Θ is a smooth cut-off function supported in $[0, R)$. Let $\Omega = \Gamma_\varepsilon^c \cap B_R$. Then $v \in L^p(\Omega)$ for all $1 < p < \infty$ and has zero boundary values. Moreover,

(3.8) $$f := \Delta v = 2\nabla u \nabla \Theta + u \Delta \Theta$$

is in $L^p(\Omega)$ for all $1 < p < \infty$ as the gradient of Θ vanishes near the tip of the cone Γ_ε.

However, by homogeneity, $v \in W^{2,p}(\Omega)$ iff $r^{\lambda-2} \in L^p(\Omega)$, i.e. $(\lambda - 2)p > -3$. As λ can be made arbitrarily small by making the cone narrower, this implies that if $p > 3/2$, we can find a bounded Lipschitz domain $\Omega \subseteq \mathbb{R}^3$ such that $v \notin W^{2,p}(\Omega)$.

We now show that v is the unique solution in a space $L^p_\alpha(\Omega)$ for some $\alpha \in (0, 2)$ to the problem

(3.9)
$$\begin{cases} \Delta u = f & \text{in } \Omega, \\ u = 0 & \text{on } \partial\Omega. \end{cases}$$

The same homogeneity consideration as above implies that we have $v \in L^p_\alpha(\Omega)$ if $\alpha \leq \frac{3}{p}$. Given $p \in (1, \infty)$, we can always find some $\alpha \in (0, 2)$ such that by Theorem 2.1.1, v is the unique solution in $L_\alpha^p(\Omega)$. In particular, this shows that there is no other solution to (3.9) in $W^{2,p}(\Omega)$.

Remarks 3.4.2. 1. Note that this does not already show that the Laplacian is not closed. It merely proves that the strong Dirichlet-Laplacian is not surjective onto $L^p(\Omega)$ for $p > 3/2$ and that the weak Dirichlet-Laplacian is not surjective onto $L^p(\Omega)$ for $p > 3$.

2. Alternatively, we can use the Strong Maximum Principle to show uniqueness. Suppose

$$\begin{cases} \Delta u = \Delta v = f & \text{in } \Omega, \\ u = v = 0 & \text{on } \partial\Omega. \end{cases}$$

Then $u - v$ is harmonic, so Weyl's Lemma (cf. [Fol76, 2.20]) implies that $u - v \in C^\infty(\Omega)$ and a result by Stampacchia (cf. [Sta65]) implies that $u - v \in C(\overline{\Omega})$. As Ω is bounded, the Maximum Principle implies $u = v$.

3. Note that this does not give uniqueness in the unbounded domain, so there could still be a $W^{2,p}$-solution in the complement of the slender cone (though obviously a trivial extension of the solution in the bounded domain would give a solution that is not in $W^{2,p}(\Omega)$).

We now use Theorem 3.4.1 to construct bounded Lipschitz domains in which the weak Dirichlet-Laplacian is not closed in $L^p(\Omega)$ for $p > 3$. Let $u(r, \omega) = r^\lambda \phi(\omega)$ be the harmonic function in the complement of the cone Γ_ε and zero on the boundary given in Theorem 3.4.1. Fix $\delta > 0$ and construct Ω_δ by rounding off the tip of the cone at a height between $\delta/2$ and δ, taking the intersection $\Gamma_\varepsilon^c \cap B_1$ and finally rounding off the corners where the cone intersects the ball as indicated in Figure 3.1. In this way we get a smooth bounded domain Ω_δ containing the set $\Gamma_\varepsilon^c \cap B_1$ and such that the Lipschitz constant of Ω_δ is uniformly bounded in δ.[2]

We solve the following Dirichlet problem.

$$(3.10) \qquad \begin{cases} \Delta u_\delta = 0 & \text{in } \Omega_\delta, \\ u_\delta = 0 & \text{on } \partial\Omega_\delta \cap \Gamma_\varepsilon, \\ u_\delta = u & \text{on } \partial\Omega_\delta \cap \Gamma_\varepsilon^c. \end{cases}$$

---

[2] In fact, for all $\delta > 0$, the Lipschitz constant of Ω_δ can be bounded by the Lipschitz constant of Γ_ε.

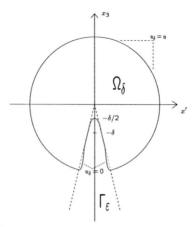

Figure 3.1: Construction of the domain Ω_δ and the function u_δ.

The solution u_δ has the following properties.

- $u_\delta \in C^2(\Omega_\delta) \cap C(\overline{\Omega_\delta})$: As u is smooth except for at 0, the boundary data is continuous and since the domain is smooth, we have that u_δ is in $C^2(\Omega_\delta)$ and continuous on the closure of the domain (see e.g. [RR93, Theorem 4.13]). In fact, as u_δ is harmonic, it is actually a C^∞-function in Ω_δ.

- On $\Gamma_\varepsilon^c \cap B_1$, by the Maximum Principle, $u_\delta \geq u$ and in all of Ω_δ, we have $u_\delta > 0$.

Our aim now is to construct a domain and a family of functions $\{w_\delta\}_{\delta>0}$ such that w_δ and Δw_δ are uniformly bounded in $L^p(\Omega)$, but ∇w_δ is not. As a first step, we show a pointwise lower bound for the gradient of u_δ.

Proposition 3.4.3. *For x in a small neighbourhood $\widetilde{\Omega}$ of the origin, we have*

$$|\nabla u_\delta(x)| \geq C \frac{u_\delta(x)}{\operatorname{dist}(x, \partial\Omega_\delta)},$$

where the constant C depends only on the Lipschitz character of Ω_δ, in particular it is independent of δ.

Proof. The proof relies on ideas and results taken from [Dah77] and [Caf87].
Let $B_{1/2}^2$ be the ball of radius $1/2$ in $\mathbb{R}^2$ and $f : B_{1/2}^2 \to \mathbb{R}$ be the function
describing the lower boundary of Ω_δ and let $\varphi(x') = f(x') + \frac{\delta}{2}$, so that
$\varphi(0) = 0.^3$ Furthermore, let

$$\tilde{u}_\delta(x) = u_\delta\left(x - \frac{\delta}{2}e_3\right).$$

For some constant M which is larger than the Lipschitz constant of φ and
$\gamma \in (0, 1/2)$, let

$$D_\gamma = \{x = (x', y) \in \mathbb{R}^3 : |x'| < \gamma, \ |y| < \gamma M, \ y > \varphi(x')\},$$

as shown in Figure 3.2.

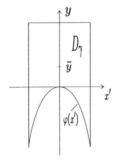

Figure 3.2: The domain D_γ.

Then $\tilde{u}_\delta$ satisfies

(3.11)
$$\begin{cases} \Delta \tilde{u}_\delta = 0 & \text{in } D_\gamma, \\ \tilde{u}_\delta \geq 0 & \text{in } D_\gamma, \\ \tilde{u}_\delta = 0 & \text{on } \partial D_\gamma \cap \{(x', \varphi(x'))\} \end{cases}$$

for any γ sufficiently small such that the boundary condition is satisfied.
By [Caf87, Lemma 5], there exists a $\gamma > 0$ such that $D_y \tilde{u}_\delta \geq 0$ in D_γ.

---

^{3}In the rest of this proof, we will always be working in this translated region.

Furthermore, by [JK82, Lemma 5.4], there exists a constant $C > 0$ and a point $(0, \tilde{y}) \in D_\gamma$ such that

$$\tilde{u}_\delta(x) \leq C\tilde{u}_\delta(0, \tilde{y}) \text{ for } x \in D_\gamma$$

where C depends only on the Lipschitz character of D_γ, in particular, it is independent of δ. Moreover, $\tilde{y}$ is independent of δ.

Now [Caf87, Lemma 4] implies that there exist constants C_1, C_2 independent of δ such that

$$0 < C_1 \leq \frac{D_y\tilde{u}_\delta(0, \tilde{y})}{\tilde{u}_\delta(0, \tilde{y})} \leq C_2.$$

Using the Harnack principle for $\tilde{u}_\delta$ and $D_y\tilde{u}_\delta$, we have that in a small neighbourhood U of $(0, \tilde{y})$,

$$0 < C_1 \leq \frac{D_y\tilde{u}_\delta(x)}{\tilde{u}_\delta(x)} \leq C_2 \text{ for } x \in U,$$

again with δ-independent constants C_1 and C_2.

Now define $\tilde{u}_{\delta,r}(x) = \tilde{u}_\delta(rx)$.[4] The functions $\tilde{u}_{\delta,r}$ satisfy the same conditions as the $\tilde{u}_\delta$, so we get

$$0 < C_1 \leq \frac{rD_y\tilde{u}_\delta(rx)}{\tilde{u}_\delta(rx)} \leq C_2 \text{ for } x \in U.$$

Furthermore, $r \cong \text{dist}(rx, \partial D_\gamma)$ for r sufficiently small.[5]

Then the desired estimate for u_δ holds with $\widetilde{\Omega} = rU - \frac{\delta}{2}e_3$ which contains a neighbourhood of the origin if r is chosen appropriately. □

We now need to use a cut-off function to obtain functions which satisfy the zero boundary condition.

Proposition 3.4.4. *Let $\Theta \equiv 1$ on $B_{1/2} \cap \Omega_\delta$ be a smooth, radially symmetric cut-off function supported in B_1 where δ is sufficiently small. Let $w_\delta = \Theta u_\delta$. Then,*

a) for $1 < p < \infty$, $w_\delta \in L^p(\Omega_\delta)$ with uniform bound in δ,

---

[4]Note that rescaling does not effect the Lipschitz constant.
[5]This easily follows from the fact that D_γ is Lipschitz by using the triangle inequality.

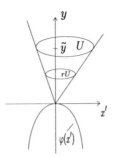

Figure 3.3: The neighbourhood U.

b) for $1 < p < \infty$, $\Delta w_\delta \in L^p(\Omega_\delta)$ with uniform bound in δ,

c) for $p > 3$, $\left(\int_{\Omega_\delta} |\nabla w_\delta|^p \right)^{\frac{1}{p}} \geq C\delta^{-\zeta}$ for some $\zeta > 0$.

Proof. a) By the Maximum Principle,

$$0 \leq u_\delta(x) \leq \max_{y \in \partial\Omega_\delta \cap \Gamma_\varepsilon} u(y) =: M$$

and M is independent of δ. As the measure of Ω_δ is uniformly bounded in δ, u_δ and therefore w_δ are in $L^p(\Omega_\delta)$ with uniform bound in δ.

b) We denote the Laplacian on the sphere by Δ_S. Then for the Laplacian of w_δ in polar coordinates we have

$$
\begin{aligned}
\Delta w_\delta &= \left(\partial_r^2 + \frac{2}{r}\partial_r + \frac{1}{r^2}\Delta_S \right) \Theta u_\delta \\
&= u_\delta \partial_r^2 \Theta + 2\partial_r\Theta\partial_r u_\delta + \Theta\partial_r^2 u_\delta + \frac{2}{r}u_\delta\partial_r\Theta + \frac{2}{r}\Theta\partial_r u_\delta + \Theta\frac{1}{r^2}\Delta_S u_\delta \\
&= u_\delta\Delta\Theta + \Theta\Delta u_\delta + 2\partial_r\Theta\partial_r u_\delta.
\end{aligned}
$$

The term with u_δ is uniformly bounded in $L^p(\Omega_\delta)$ by the first part of the proof, and $\Delta u_\delta = 0$. It remains to show that the term $\partial_r\Theta\partial_r u_\delta$ is uniformly bounded. To do this we adapt a standard procedure for interior gradient estimates of harmonic functions to our situation. The aim is to construct a subharmonic function involving $\partial_r u_\delta$ and to apply the Maximum Principle.

We assume w.l.o.g. that the support of $\nabla\Theta$ is contained within an annulus A_0 such that the intersection of A_0 with the boundary of Ω_δ is contained in the boundary of the cone Γ_ε. Let $A = A_0 \cap \Omega_\delta$ (cf. Figure 3.4). By doing this, we have that, along $\partial A \cap \partial \Gamma_\varepsilon$, $\partial_r u_\delta$ is the tangential derivative of u_δ. Due to the Dirichlet boundary conditions, it then vanishes on this part of ∂A.

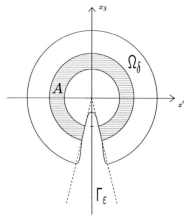

Figure 3.4: The ring A containing supp($\nabla\Theta$).

Now set $v(x) = r\partial_r u_\delta(x)$. Since $r\partial_r f = \sum_{j=1}^n x_j \partial_j f$, we have

$$
\begin{aligned}
\Delta v &= \Delta\left(\sum_{j=1}^n x_j \partial_j u_\delta\right) \\
&= \sum_{i,j=1}^n \left(x_j \partial_j \partial_i^2 u_\delta + 2\delta_{ij}\partial_i\partial_j u_\delta\right) \\
&= \sum_{j=1}^n x_j \partial_j \Delta u_\delta + 2\Delta u_\delta = 0,
\end{aligned}
$$

so v is harmonic. Note that, as a harmonic function, $u_\delta \in C^\infty(\Omega_\delta)$, so $v \in C^\infty(\Omega_\delta)$ and all partial derivatives exist and commute with each other. Moreover, $\Delta v^2 = 2v\Delta v + 2|\nabla v|^2 = 2|\nabla v|^2 \geq 0$. Therefore, v^2 is subharmonic.

We could apply the Maximum Principle to v, however we have no knowledge of the behaviour of v on the boundary of A. Instead, we introduce another cut-off function $\zeta \in C_c^\infty(\mathbb{R}^3)$ such that $\zeta \equiv 1$ on $\mathrm{supp}(\nabla\Theta)$ and $\zeta \equiv 0$ on A^c. Then

$$\Delta(\zeta^2 v^2) = \left(2\zeta\Delta\zeta + 2|\nabla\zeta|^2\right) v^2 + 8\zeta v \sum_{i=1}^n \partial_i\zeta\partial_i v + 2\zeta^2|\nabla v|^2.$$

Our aim is now to get rid of the terms depending on derivatives of v. Setting $a_i := 2v\partial_i\zeta$ and $b_i = \zeta\partial_i v$, from $b_i^2 + 2a_i b_i \geq -a_i^2$, we get

$$\zeta^2|\nabla v|^2 + 4\zeta v \sum_{i=1}^n \partial_i\zeta\partial_i v \geq -4|\nabla\zeta|^2 v^2$$

and so

$$\Delta(\zeta^2 v^2) \geq \left(2\zeta\Delta\zeta + 2|\nabla\zeta|^2\right) v^2 - 8|\nabla\zeta|^2 v^2 \geq -Cv^2$$

for some $C \geq 0$ which depends on ζ but is independent of δ.

We can estimate $v^2 = \left(\sum_{i=1}^n x_i\partial_i u_\delta\right)^2 \leq C'|\nabla u_\delta|^2$ and u_δ^2 is subharmonic with $\Delta u_\delta^2 = 2|\nabla u_\delta|^2$. For some sufficiently large constant α, the function $w = \zeta^2 v^2 + \alpha u_\delta^2$ is subharmonic, as

$$\Delta w = \Delta(\zeta^2 v^2) + \Delta(\alpha u_\delta^2) \geq -Cv^2 + 2\alpha|\nabla u_\delta|^2 \geq (-C + 2\alpha/C')v^2 \geq 0.$$

By the Maximum Principle (see e.g. [RR93, Theorem 4.3]), we have $\max_A w \leq \max_{\partial A} w$. However, by construction the part $\zeta^2 v^2$ vanishes on ∂A as ζ is zero on the two arcs and v is zero on the radial part of the boundary. Therefore, by part a), we get

$$\max_{\mathrm{supp}(\nabla\Theta)} v^2 \leq \max_A w \leq \max_{\partial A} \alpha u_\delta^2 \leq \alpha M^2,$$

i.e. $\partial_r u_\delta$ is bounded independently of δ on $\mathrm{supp}(\nabla\Theta)$ which gives a uniform bound on the term $\partial_r\Theta\partial_r u_\delta$ in $L^p(\Omega_\delta)$.

c) We consider w_δ in a shell above the tip of the cone

$$S_\delta = \{(r,\varphi,\theta) : r \in (\delta/2,\delta), \varphi \in [0,2\pi), \theta \in [0,\alpha)\}$$

for some small α as shown in Figure 3.5. Assume δ is sufficiently small so

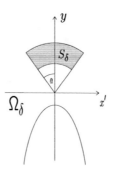

Figure 3.5: The shell S_δ.

that $\Theta \equiv 1$ in S_δ and such that the conclusion of Proposition 3.4.3 holds for all $x \in S_\delta$. Then for $x \in S_\delta$ we have

$$|\nabla w_\delta(x)| = |\nabla u_\delta(x)| \geq C\frac{u_\delta(x)}{\delta} \geq \frac{C}{\delta}u(x) = \frac{C}{\delta}r^\lambda\phi(\omega).$$

Note that, by the Maximum Principle, $u > 0$ in S_δ, so $\phi > 0$ and by compactness, $\phi > c > 0$ in S_δ. Moreover, as ϕ only depends on the angle variable ω, this lower estimate on ϕ is independent of δ. Therefore,

$$\left(\int_{\Omega_\delta} |\nabla w_\delta|^p\right)^{\frac{1}{p}} \geq \frac{C}{\delta}\delta^\lambda(\mu(S_\delta))^{1/p} = C\delta^{\lambda-1+3/p}.$$

So for any $p > 3$, we can make λ sufficiently small to get a negative power of δ.[6] $\square$

We now use this construction to prove the following negative results.

Theorem 3.4.5. *Let $p > 3$ and let $\Delta_{p,w}^D$ be the weak Dirichlet-Laplacian defined in Definition 3.1.1. Then there exist*

  a) an unbounded smooth domain $\Omega \subset \mathbb{R}^3$ and

---

[6]Note that the constant C depends on the Lipschitz constant of Ω_δ which in turn depends on the angle of the cone Γ_ϵ on which the exponent λ also depends. Therefore, C may depend on λ. However, here we only need that C is independent of δ.

b) a bounded Lipschitz domain $\Omega \subset \mathbb{R}^3$,

such that $\Delta^D_{p,w}$ is not closed in $L^p(\Omega)$.

Proof.   We start with the unbounded case. Let $\{\delta_j\}_{j \in \mathbb{N}} \subseteq \mathbb{R}_+$ such that $\delta_j \to 0$. Construct the domains Ω_{δ_j} and functions w_{δ_j} as before. Now, put infinitely many cones in a row with tips at $x_j = (x'_j, 0)$ and with sufficient distance between them such that the the sets $\Omega_{\delta_j} + x_j$ are disjoint. Let Ω be a smooth domain above the graph of a function φ containing the sets $\Omega_{\delta_j} + x_j$, where, in a neighbourhood of x'_j, φ describes the lower boundary of $\Omega_{\delta_j} + x_j$ (see Figure 3.6). For $x \in \Omega$ consider the function

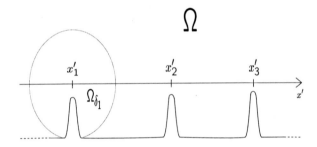

Figure 3.6: The unbounded domain Ω.

$$w(x) = \sum_{j=0}^{\infty} 2^{-j} w_{\delta_j}(x - x_j),$$

where we extend w_{δ_j} to Ω by zero outside Ω_{δ_j}. By Proposition 3.4.4, w and Δw are in $L^p(\Omega)$. We can approximate w in $L^p(\Omega)$ by

$$w_N(x) = \sum_{j=0}^{N} 2^{-j} w_{\delta_j}(x - x_j).$$

Then $w_N \in D(\Delta_{p,w}^D)$ and $\Delta w_N(x) \to \sum_{j=0}^{\infty} 2^{-j} \Delta w_{\delta_j}(x - x_j)$ in $L^p(\Omega)$. However, for $p > 3$, using that the functions w_{δ_j} have disjoint support, we get

$$\left(\int_\Omega |\nabla w|^p \right)^{\frac{1}{p}} = \sum_{j=0}^{\infty} 2^{-j} \left(\int_{\Omega_{\delta_j}} |\nabla w_{\delta_j}|^p \right)^{\frac{1}{p}} \geq \sum_{j=0}^{\infty} 2^{-j} \delta_j^{-\varsigma}.$$

Let $\delta_j \to 0$ sufficiently fast such that the sum diverges. Then $w \notin D(\Delta_{p,w}^D)$, so the operator is not closed.

Next, we consider the case of a bounded domain by using a scaling argument. Let φ be a Lipschitz function describing a cone with the tip at $(0, 0, 1)$. Let φ_δ be a smooth function gained from φ by rounding off the tip of the cone between a height of $1 - \delta$ and $1 - \delta/2$ and such that $\|\varphi_\delta\|_\infty \leq 1$. Construct w_δ as before, allowing for the shift of the tip of the cone away

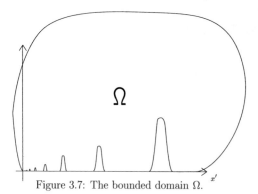

Figure 3.7: The bounded domain Ω.

from the origin. For a sequence $(r_j) \subseteq \mathbb{R}_+$, define $w_{\delta,j}(x) = w_\delta(r_j^{-1}x)$ for $x \in \Omega_{\delta,j} = \{(x_1, x_2, x_3) \in \mathbb{R}^3 : x_3 > r_j \varphi_\delta(r_j^{-1}x_1, r_j^{-1}x_2)\}$. The scaled domains all have the same Lipschitz constant. $w_{\delta,j}$ has the following properties.

$$\int_{\Omega_{\delta,j}} |w_{\delta,j}(x)|^p = r_j^3 \int_{\Omega_\delta} |w_\delta|^p \leq C r_j^3,$$

$$\int_{\Omega_{\delta,j}} |\Delta w_{\delta,j}(x)|^p = r_j^{3-2p} \int_{\Omega_\delta} |\Delta w_\delta|^p \leq C r_j^{3-2p},$$

$$\int_{\Omega_{\delta,j}} |\nabla w_{\delta,j}(x)|^p = r_j^{3-p} \int_{\Omega_\delta} |\nabla w_\delta|^p \geq C r_j^{3-p} \delta^{-\varsigma}.$$

Let $x_j = (2^{-j}, 0, 0)$, $r_j = 8^{-j}$ and

$$w(x) = \sum_{j=0}^{\infty} 8^{-3j(1-1/p)} w_{\delta_j, j}(x - x_j).$$

Then w and Δw are in $L^p(\Omega)$ where Ω is a bounded domain containing the sets $\Omega_{\delta_j, j} + x_j$ with its lower boundary given by $\sum_{j=0}^{\infty} r_j \varphi_\delta(r_j^{-1}(x - x_j))^7$ as in Figure 3.7. Now let $\delta_j \to 0$ sufficiently fast, so that $\nabla w \notin L^p(\Omega)$ for $p > 3$. By approximating w by the partial sums, we see that $\Delta_{p,w}^D$ is not closed on $L^p(\Omega)$.   □

In higher dimensions, we get the same result.

Corollary 3.4.6. *Let $p > 3$ and $n \geq 3$. Then there exist*

  a) an unbounded smooth domain $\Omega \subseteq \mathbb{R}^n$ and

  b) a bounded Lipschitz domain $\Omega \subseteq \mathbb{R}^n$,

such that $\Delta_{p,w}^D$ is not closed in $L^p(\Omega)$.

Proof. We can use the same construction as in $\mathbb{R}^3$ by setting $\mathbb{R}^n = \mathbb{R}^3 \times \mathbb{R}^{n-3}$. Then let $\widetilde{u}(r, \omega, x) = u(r, \omega)$ for $r \in \mathbb{R}_+$, $\omega \in \mathbb{S}^2 \cap \Gamma_\varepsilon^c$ and $x \in \mathbb{R}^{n-3}$ with Γ_ε and u as in Theorem 3.4.1 and $\mathbb{S}^2$ the unit sphere in $\mathbb{R}^3$. Then we can construct smooth bounded domains $\Omega_\delta \subseteq \mathbb{R}^n$ and functions u_δ and w_δ as in the case of three dimensions.   □

The same examples also show negative results in the case of the strong Dirichlet-Laplacian.

Corollary 3.4.7. *Let $p > 3$, $n \geq 3$ and let $\Delta_{p,s}^D$ be defined as in Definition 3.1.2. Then there exist*

  a) an unbounded smooth domain $\Omega \subseteq \mathbb{R}^n$ and

  b) a bounded Lipschitz domain $\Omega \subseteq \mathbb{R}^n$,

---

[7]This is necessary, as it is not possible to smoothly continue the functions $w_{\delta,j}$ by zero across this part of the boundary of $\Omega_{\delta,j}$. Also note that Ω is Lipschitz but not C^1 as the derivatives of the function describing the boundary do not converge at zero.

such that $\Delta_{p,s}^D$ *is not closed in* $L^p(\Omega)$.

Remarks 3.4.8. Obviously, for the strong Dirichlet-Laplacian this still leaves many unresolved cases. In fact, especially in light of the counterexample by Dahlberg for Laplace's equation stated in Section 2.1, one might expect that for any $1 < p < \infty$ there should be a Lipschitz domain Ω where $\Delta_{p,s}^D$ is not closed in $L^p(\Omega)$. The approach in the counterexample given here however relies on getting a lower bound on the gradient of the solution to the Dirichlet problem and does not carry over to higher order derivatives of the solution or other boundary conditions.

Chapter 4

Operators in non-divergence form with L^∞-coefficients

In this chapter we use the results we have gained for the Laplacian in Chapter 3 to consider elliptic and parabolic equations in convex bounded domains $\Omega \subseteq \mathbb{R}^n$ for second order differential operators in non-divergence form with coefficients $a_{ij} \in L^\infty(\Omega)$. More precisely, we are interested in the L^p-theory of the elliptic problem $\mathcal{A}u = f$ with homogeneous Dirichlet data and in the parabolic problem $u_t - \mathcal{A}u = f$ with homogeneous Dirichlet and initial data, where

$$\mathcal{A} := \sum_{i,j=1}^{n} a_{ij}(x,t) D_{ij}$$

is a second order differential operator with coefficients $a_{ij} \in L^\infty(\Omega \times \mathbb{R}_+, \mathbb{C})$. Whereas the autonomous situation for operators in divergence form, i.e. $\mathcal{A}u = \text{div}(a_{ij}(x)\nabla u)$, is fairly well understood (see [Aus04] for the latest results), the situation is less clear for operators in non-divergence form. If Ω is a bounded domain with smooth boundary, i.e. of class $C^{2+\alpha}$ and the coefficients a_{ij} are continuous on $\overline{\Omega}$, then by classical results of Agmon, Douglis, Nirenberg [ADN59] and Ladyzhenskaya, Solonnikov and Uralceva [LSU68] a very satisfactory L^p-regularity theory in the elliptic and the parabolic case is known for all p satisfying $1 < p < \infty$.

On the other hand, there exist examples of strongly elliptic operators $\mathcal{A}$ in non-divergence form, i.e. $\mathcal{A} := \sum_{i,j=1}^{n} a_{ij}(x) D_{ij}$ with $a_{ij} \in L^\infty(\Omega)$ such that for $f \in L^2(\Omega)$ the elliptic Dirichlet problem does not admit a solution $u \in H^2(\Omega) \cap H_0^1(\Omega)$. We will take a closer look at such an example in Section 4.1.

Starting from this situation, elliptic operators $\mathcal{A}$ with $a_{ij} \in L^\infty(\Omega)$ satisfying a so-called Cordes condition [Cor56] were investigated in detail on smooth domains by Campanato [Cam67] and Talenti [Tal65]. They proved existence and uniqueness results as well as regularity properties in the L^p-context for p close to 2. For a systematic treatment and a survey of these results, we refer to Maugeri, Palagachev and Softova [MPS00].

Our main aim here is to show that results of the above type hold true also for domains with non-smooth boundary, more precisely for bounded and convex domains Ω and for $p \in (p_0, 2]$ where p_0 is sufficiently close to 2. The analysis of the elliptic problem relies on the results on the Laplacian in convex domains given in Section 2.1.

In Section 4.3, we solve the parabolic problem for operators with L^∞-coefficients subject to a parabolic Cordes condition. To do this, we use the maximal regularity for the strong Laplacian shown in Theorem 3.3.2. Not only are we able to treat variable coefficient operators depending on x and t, but in addition to Shen's results for elliptic systems with real constant coefficients [She95], we are able to give a precise description of the domain of the operators and show maximal regularity estimates in convex bounded domains.

In two space dimensions, i.e. for $n = 2$, strong ellipticity of $\mathcal{A}$ is sufficient for $\mathcal{A}$ to satisfy the Cordes condition for elliptic equations. For parabolic equations, unfortunately, an additional assumption on the coefficients is needed (see (4.16)). It follows from our approach that many strongly elliptic operators in two dimensions generate analytic semigroups on $L^p(\Omega)$ provided $p \in (p_0, 2]$ and p_0 is sufficiently close to 2. This is remarkable, since results of this type cannot be obtained by the usual localisation technique.

4.1 A counterexample

That some extra assumptions on the coefficients of a non-divergence form operator besides ellipticity are necessary to prove unique solvability of the Dirichlet problem can be seen by the following counterexample which first appeared in Talenti [Tal65]. Note that the example is in a smooth convex domain Ω and even in the Hilbert space setting $L^2(\Omega)$.

Let $n > 2$, Ω be the unit ball in $\mathbb{R}^n$ and set

$$D(A) := H^2(\Omega) \cap H_0^1(\Omega),$$
$$A := \sum_{i,j=1}^n \left(\frac{x_i x_j}{|x|^2}(1 - cn) + \delta_{ij}c \right) D_{ij},$$

for $0 < c < \frac{n-2}{n(n-1)}$. A is strongly elliptic, as

$$
\begin{aligned}
\sum_{i,j=1}^n a_{ij}\xi_i\xi_j &= \sum_{i,j=1}^n \frac{x_i x_j}{|x|^2}(1 - cn)\xi_i\xi_j + c|\xi|^2 \\
&= \frac{\langle x, \xi \rangle^2}{|x|^2}(1 - cn) + c|\xi|^2 \\
&\geq c|\xi|^2.
\end{aligned}
$$

However, we will now show that $A : H^2(\Omega) \cap H_0^1(\Omega) \to L^2(\Omega)$ is not an isomorphism.

In Talenti [Tal65] it was proved that the operator $A^{-1} : L^2(\Omega) \to H^2(\Omega) \cap H_0^1(\Omega)$ cannot be bounded. Here, we will show by a simple calculation that the operator A is not injective.

Let $u(x) := |x|^\alpha - 1$, $\alpha \in \mathbb{R}$. Obviously u satisfies the boundary condition for all real α. We have

$$
\begin{aligned}
D_i u &= \alpha x_i |x|^{\alpha-2}, \\
D_{ij} u &= \delta_{ij}\alpha|x|^{\alpha-2} + \alpha(\alpha-2)x_i x_j |x|^{\alpha-4}.
\end{aligned}
$$

The second order derivatives are homogeneous of degree $\alpha - 2$ and switching to polar coordinates we see that $u \in H^2(\Omega)$ iff $\alpha > 2 - n/2$. Now,

$$Au = \sum_{i,j=1}^n \left(\frac{x_i x_j}{|x|^2}(1 - cn) + \delta_{ij}c \right) \left(\delta_{ij}\alpha|x|^{\alpha-2} + \alpha(\alpha-2)x_i x_j |x|^{\alpha-4} \right)$$

$$= (1 - cn)(\alpha|x|^{\alpha-2} + \alpha(\alpha-2)|x|^{\alpha-2}) + cn\alpha|x|^{\alpha-2} + c\alpha(\alpha-2)|x|^{\alpha-2}$$
$$= \alpha|x|^{\alpha-2}(1 + (\alpha-2)(1 - cn + c)).$$

Choosing $\alpha = 2 - \frac{1}{1-c(n-1)}$ for $0 < c < \frac{n-2}{n(n-1)}$, we get that $u \in D(A)$ and $Au = 0$. If $c \neq \frac{1}{2(n-1)}$, we have $\alpha \neq 0$ and then obviously $u(x) \equiv 0$ is a second solution. This proves that A is not injective.

4.2 Elliptic equations

Let Ω be a bounded, convex domain in $\mathbb{R}^n$. Consider the operator

$$\mathcal{A}u(x) := \sum_{i,j=1}^{n} a_{ij}(x)D_{ij}u(x).$$

We now introduce the conditions on the coefficients a_{ij} that guarantee existence and uniqueness of the solution to the elliptic problem. Assume that the coefficients $a_{ij} \in L^\infty(\Omega)$ are complex-valued, that

(4.1) $$\sum_{i=1}^{n} \text{Re } a_{ii}(x) \neq 0 \text{ a.e. } x \in \Omega,$$

and that they satisfy a *Cordes condition*, i.e. there exists $\varepsilon > 0$ such that

(4.2) $$\frac{\sum_{i,j=1}^{n} |a_{ij}(x)|^2}{(\sum_{i=1}^{n} \text{Re } a_{ii}(x))^2} \leq \frac{1}{n - 1 + \varepsilon} \text{ a.e. } x \in \Omega.$$

Moreover, we assume that

(4.3) $$\alpha(x) := \frac{\sum_{i=1}^{n} \text{Re } a_{ii}(x)}{\sum_{i,j=1}^{n} |a_{ij}(x)|^2} \in L^\infty(\Omega).$$

Note that (4.1) and (4.3) are satisfied whenever $\mathcal{A}$ has real-valued coefficients and satisfies the ellipticity condition

$$\sum_{i,j=1}^{n} a_{ij}(x)\xi_i\xi_j \geq \mu|\xi|^2, \text{ a.e. } x \in \Omega \text{ and all } \xi \in \mathbb{R}^n,$$

for some $\mu > 0$, since then we have that the coefficients a_{ii} satisfy $a_{ii} \geq \mu$. The Laplacian satisfies (4.2) for any $\varepsilon \in (0,1]$. The quotient $\frac{\sum_{i,j=1}^{n} |a_{ij}(x)|^2}{(\sum_{i=1}^{n} \text{Re } a_{ii}(x))^2}$ gives a measure of how far the coefficients of $\mathcal{A}$ vary from those of the Laplacian while (4.3) guarantees that not all coefficients can simultaneously tend to zero.

Example 4.2.1. *Let $b \in L^\infty(\Omega; \mathbb{C})$. Then the operator $\mathcal{A} = b(x)\Delta^1$ satisfies the Cordes condition with $\varepsilon > 0$ if b takes values in the double sector Σ given by*

$$\left| \frac{\operatorname{Im} b(x)}{\operatorname{Re} b(x)} \right| \leq \sqrt{\frac{1 - \varepsilon}{n - 1 + \varepsilon}} \quad a.e. \ x \in \Omega.$$

In other words, whenever

$$\left| \frac{\operatorname{Im} b(x)}{\operatorname{Re} b(x)} \right| \leq \sqrt{\frac{1}{n - 1}} \quad a.e. \ x \in \Omega,$$

we can find an $\varepsilon > 0$ such that $b(x)\Delta$ satisfies the Cordes condition (4.2).

Moreover, (4.3) is satisfied as long as b is bounded away from the origin.

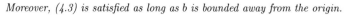

Proof. Let

$$c(x) := \operatorname{Re} b(x) \quad \text{and} \quad \alpha(x) := \frac{\operatorname{Im} b(x)}{\operatorname{Re} b(x)}.$$

Then $b(x) = (1 + \alpha(x)i)c(x)$, and the Cordes condition is satisfied if

$$\frac{nc(x)^2(1 + \alpha^2(x))}{n^2 c(x)^2} \leq \frac{1}{n - 1 + \varepsilon}.$$

This is the case if $|\alpha(x)| \leq \sqrt{\frac{1-\varepsilon}{n-1+\varepsilon}}$. $\qquad\square$

Our aim is to solve the following problem. Let $1 < p \leq 2$ and let A be the operator defined by

$$(4.4) \qquad\qquad Au \ := \ \mathcal{A}u,$$

$$(4.5) \qquad\qquad D(A) \ := \ W^{2,p}(\Omega) \cap W_0^{1,p}(\Omega).$$

Given $f \in L^p(\Omega)$, find $u \in D(A)$ such that

$$(4.6) \qquad\qquad Au = f \quad \text{in } \Omega.$$

Remarks 4.2.2. Note that from Theorem 3.3.2, we know that the Laplacian with domain $D(A)$ is a closed operator in $L^p(\Omega)$. In the following, we will

---

[1] Here, the Laplacian Δ simply stands for the sum of all pure second derivatives without any consideration of the domain.

see that if $\mathcal{A}$ satisfies the Cordes condition and p is close to 2, then also A is a closed operator (cf. Theorem 4.2.9). In general, this need not be the case.

As a first step, we introduce the following Banach space V_p.

Proposition 4.2.3. *Let $1 < p \leq 2$ and $V_p := W^{2,p}(\Omega) \cap W_0^{1,p}(\Omega)$ equipped with the norm*

$$\|u\|_{V_p} := \left(\int_\Omega \left(\sum_{i,j=1}^n |D_{ij}u|^2 \right)^{p/2} \right)^{1/p}.$$

Then $(V_p, \|\cdot\|_{V_p})$ is a Banach space.

Proof. We show that in V_p the norm $\|\cdot\|_{V_p}$ is equivalent to the usual Sobolev norm in $W^{2,p}(\Omega)$. This will then obviously imply the statement. It is clear that for $u \in V_p$, we have $\|u\|_{V_p} \leq \|u\|_{W^{2,p}}$. On the other hand, for any $u \in V_p$, Theorem 2.1.2 implies

$$\|u\|_{W^{2,p}} \leq C \|\Delta u\|_p \leq C \|u\|_{V_p}.$$

Therefore, as a closed subspace of $W^{2,p}(\Omega)$, V_p is a Banach space. □

Remarks 4.2.4. Observe that this result relies on Dirichlet boundary conditions. If we imposed Neumann boundary conditions, the solution to Laplace's equation would only be defined up to a constant. Therefore, we could never expect to estimate the full $W^{2,p}$-norm by the second derivatives.

4.2.1 The elliptic case for $p = 2$

We first consider the problem (4.6) for the case $p = 2$.

Proposition 4.2.5. *The norm in V_2 is bounded by the L^2-norm of the Laplacian, i.e.*

$$\int_\Omega \sum_{i,j=1}^n |D_{ij}u|^2 \leq \int_\Omega |\Delta u|^2.$$

Proof. This is due to the convexity of the domain Ω. See [Fro93] or [Gri85, Section 3.1]. $\qquad\square$

The idea is now to view A as a small perturbation of the Laplacian where the size of the perturbation is measured in the Cordes condition. Then we solve the problem $Au = f$ by solving $\Delta u = (\Delta - A)w + f$ for $w \in V_2$ and searching for a fixed point of this equation. This gives us our first result.

Proposition 4.2.6. *Let Ω be a bounded convex domain in $\mathbb{R}^n$. Let A be defined as in (4.4) and (4.5) and assume that (4.1), (4.2) and (4.3) are satisfied. Given $f \in L^2(\Omega)$, there is a unique $u \in H^2(\Omega) \cap H_0^1(\Omega)$ satisfying $Au = f$ in Ω.*

Proof. Let α be defined as in (4.3). Denote by $\Phi : V_2 \to V_2$ the map defined by $u = \Phi w$, where, for given $w \in V_2$, u is the unique solution of

$$(4.7) \qquad \begin{cases} \Delta u = \alpha f + \Delta w - \alpha Aw & \text{in } \Omega, \\ u = 0 & \text{on } \partial\Omega. \end{cases}$$

As, using assumption (4.3), the right hand side above is in $L^2(\Omega)$, by Theorem 2.1.2, there exists a unique function $u \in V_2$ satisfying (4.7). Now let $w, \overline{w} \in V_2$ and $u = \Phi(w), \overline{u} = \Phi(\overline{w})$. Then, using Proposition 4.2.5, we obtain

$$\begin{aligned} \|\Phi w - \Phi \overline{w}\|_{V_2}^2 &= \|u - \overline{u}\|_{V_2}^2 \\ &\leq \|\Delta(u - \overline{u})\|_{L^2}^2 \\ &= \|\Delta(w - \overline{w}) - \alpha A(w - \overline{w})\|_{L^2}^2 \\ &= \int_\Omega \left|\sum_{i,j=1}^n (\delta_{ij} - \alpha a_{ij})D_{ij}(w - \overline{w})\right|^2 \\ &\leq \int_\Omega \left(\sum_{i,j=1}^n |\delta_{ij} - \alpha a_{ij}|^2\right)\left(\sum_{i,j=1}^n |D_{ij}(w - \overline{w})|^2\right). \end{aligned}$$

Since, using our definition of α and (4.2),

$$\begin{aligned} \sum_{i,j=1}^n |\delta_{ij} - \alpha a_{ij}|^2 &= \sum_{i,j=1}^n \left(\delta_{ij}^2 - 2\alpha\delta_{ij} \operatorname{Re} a_{ij} + \alpha^2|a_{ij}|^2\right) \\ &= n - 2\alpha \sum_{i=1}^n \operatorname{Re} a_{ii} + \alpha^2 \sum_{i,j=1}^n |a_{ij}|^2 \end{aligned}$$

$$\begin{aligned}
&= n - \frac{\left(\sum_{i=1}^n \mathrm{Re}\, a_{ii}\right)^2}{\sum_{i,j=1}^n |a_{ij}|^2} \\
&\leq 1 - \varepsilon,
\end{aligned}$$

we obtain

$$\|\Phi w - \Phi \overline{w}\|_{V_2}^2 \leq (1 - \varepsilon) \|w - \overline{w}\|_{V_2}^2 .$$

By the Banach Fixed Point Theorem, we get a unique solution $u \in V_2$ such that $\Phi u = u$, which is the desired solution to $Au = f$. □

4.2.2   The elliptic case for $p \neq 2$

Our aim is to find a solution to $Au = f$ with $f \in L^p(\Omega)$ in the space $V_p := W^{2,p}(\Omega) \cap W_0^{1,p}(\Omega)$. We first need to transfer Proposition 4.2.5 to the L^p-context.

Proposition 4.2.7. *Let $1 < p \leq 2$. Given $f \in L^p(\Omega)$, there exists a constant C_p and a unique $u \in V_p$ satisfying $\Delta u = f$ with $\|u\|_{V_p} \leq C_p \|f\|_p$. Furthermore, for any $\varepsilon \in (0,1)$, there exists $\delta > 0$ such that for $p \in (2 - \delta, 2]$, the constant C_p satisfies the estimate $C_p < (1 - \varepsilon)^{-1/2}$.*

Proof. Once more using Theorem 2.1.2, we may define a bounded operator T on $L^p(\Omega)$ such that

$$Tf := \left(\sum_{i,j=1}^n |D_{ij}u|^2 \right)^{1/2}$$

where u is the unique solution of $\Delta u = f$ in V_p. Thus $\|Tf\|_p \leq C_p \|f\|_p$ for some C_p. By Proposition 4.2.5, $C_2 = 1$. Hence, by the Riesz-Thorin Interpolation Theorem, C_p is close to 1 for p near 2. □

Remarks 4.2.8. There exist a bounded convex domain Ω and an $f \in C^\infty(\overline{\Omega})$ such that for any $p > 2$ the solution to

$$\begin{cases} \Delta u = f & \text{in } \Omega \\ \quad u = 0 & \text{on } \partial\Omega \end{cases}$$

is not in $W^{2,p}(\Omega)$ (cf. [Fro93]). Therefore, an extension of the results to $p > 2$ is impossible.

Our main result for the elliptic situation is the following:

Theorem 4.2.9. *Let Ω be a bounded convex domain in $\mathbb{R}^n$ and let A be an operator defined as in (4.4) and (4.5) satisfying (4.1), (4.2) for some $\varepsilon > 0$ and (4.3). Then there exists $\delta > 0$ such that for $p \in (2 - \delta, 2]$ and $f \in L^p(\Omega)$, there is a unique $u \in V_p$ satisfying $Au = f$ in Ω and we have the estimate*

$$(4.8) \qquad \|u\|_{V_p} \leq C \|f\|_p.$$

Proof. Define α and Φ as in the proof of Theorem 4.2.6. Using Proposition 4.2.7, a similar calculation as in the proof of Theorem 4.2.6 shows that Φ is a contraction on V_p. The Banach Fixed Point Theorem then yields the desired solution and the estimate follows from the closed graph theorem. $\qquad\square$

Remarks 4.2.10. The estimate (4.8) implies that A is a closed operator, as then $\|\cdot\|_{V_p}$ and $\|A\cdot\|_p$ are equivalent norms on V_p.

The case of two dimensions is of particular interest, as, in this case, strong ellipticity and symmetry are sufficient to guarantee the Cordes condition.

Corollary 4.2.11. *Let Ω be a bounded convex domain in $\mathbb{R}^2$. Assume that the coefficients of the operator A defined as in (4.4) and (4.5) satisfy $a_{ij} \in L^\infty(\Omega, \mathbb{R})$ and that $a_{12} = a_{21}$. Assume further that A is a strongly elliptic operator, i.e. there exists a constant $\mu > 0$ such that for all $\xi \in \mathbb{R}^n$*

$$(4.9) \qquad \sum_{i,j=1}^{n} a_{ij}(x)\xi_i\xi_j \geq \mu|\xi|^2 \ a.e. \ x \in \Omega.$$

Then there exists $\delta > 0$ such that for $p \in (2 - \delta, 2]$ and $f \in L^p(\Omega)$, there is a unique $u \in V_p = W^{2,p}(\Omega) \cap W_0^{1,p}(\Omega)$ satisfying $Au = f$ in Ω.

Proof. Obviously, (4.1) and (4.3) are satisfied as $a_{11}(x), a_{22}(x) \geq \mu$ for a.e. $x \in \Omega$. By the theorem, it remains to be shown that (4.2) holds for some $\varepsilon > 0$. For this, we need that $(1+\varepsilon)(a_{11}^2 + 2a_{12}^2 + a_{22}^2) \leq a_{11}^2 + 2a_{11}a_{22} + a_{22}^2$. Due to the boundedness of the coefficients, it is enough to show that $a_{12}^2 < a_{11}a_{22}$. This follows from the ellipticity condition (4.9) as is easily seen by setting $\xi_1 = \sqrt{a_{22}}$ and $\xi_2 = \pm\sqrt{a_{11}}$. $\qquad\square$

Remarks 4.2.12. a) Strong ellipticity without the symmetry condition $a_{12} = a_{21}$ is not sufficient for the Cordes condition to be satisfied. This can be seen from the example

$$A = \begin{pmatrix} 10 & 9 \\ 0 & 3 \end{pmatrix}.$$

b) Furthermore, we cannot allow for complex-valued coefficients by replacing strong ellipticity by parameter ellipticity, i.e. that

$$A(\xi) = \sum_{i,j=1}^{n} a_{ij}(x)\xi_i\xi_j \in \Sigma_\varphi$$

for some $\varphi \in (0, \pi]$, all $\xi \in \mathbb{R}^n \setminus \{0\}$ and a.e. $x \in \Omega$. In this case,

$$A = \begin{pmatrix} 1 & i \\ -i & 1 \end{pmatrix}$$

shows that this is not a sufficient condition for (4.2) to be satisfied for any $\varepsilon > 0$.

c) If we require $a_{12} = a_{21}$, then

$$A = \begin{pmatrix} \alpha & 1+i \\ 1+i & \alpha \end{pmatrix}$$

is a parameter elliptic operator for $\alpha > 1$, but fails the Cordes condition for $\alpha < \sqrt{2}$.

4.3   Non-autonomous parabolic equations

In this section we show that similar results to those for elliptic equations also hold in the parabolic case. Let Ω be a bounded convex domain in $\mathbb{R}^n$, $J := [0, T]$, or $J := [0, \infty)$ and $Q := \Omega \times J$. Consider the operator

$$\mathcal{A}(t) := \sum_{i,j=1}^{n} a_{ij}(x, t) D_{ij}$$

with $a_{ij} \in L^\infty(Q, \mathbb{C})$. We assume that $\mathcal{A}$ satisfies a *modified parabolic Cordes condition*. By this we mean that there exists $\lambda > 0$ such that

$$(4.10) \qquad \sum_{i=1}^{n} \operatorname{Re} a_{ii}(x, t) + \lambda^{-1} \neq 0 \text{ a.e. } (x, t) \in Q,$$

and there exists $\varepsilon > 0$ such that

(4.11) $$\frac{\sum_{i,j=1}^{n} |a_{ij}(x,t)|^2 + \lambda^{-2}}{(\sum_{i=1}^{n} \operatorname{Re} a_{ii}(x,t) + \lambda^{-1})^2} \leq \frac{1}{n+\varepsilon} \quad \text{a.e. } (x,t) \in Q.$$

For a similar version of the parabolic Cordes condition we refer to [MPS00].

Example 4.3.1. *As in the example in the elliptic case, let the Laplacian Δ simply denote the sum of all pure second derivatives without any considera- tion of the domain. Let $b \in \mathbb{C}$, $\operatorname{Re} b > 0$.*

Then if b lies inside the sector Σ given by

$$\left|\frac{\operatorname{Im} b}{\operatorname{Re} b}\right| < \sqrt{\frac{1}{n-1}},$$

we can find $\lambda > 0$ and $\varepsilon > 0$ such that the operator $\mathcal{A} = b\Delta$ satisfies the modified par- abolic Cordes condition with λ and ε.

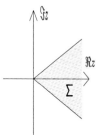

Proof. Let $\alpha = \frac{\operatorname{Im} b}{\operatorname{Re} b}$. Then the Cordes condition for $b\Delta$ is given by

$$\frac{n(\operatorname{Re} b)^2(1+\alpha^2) + \lambda^{-2}}{(n \operatorname{Re} b + \lambda^{-1})^2} \leq \frac{1}{n+\varepsilon}.$$

The roots of the equation

$$\frac{n(\operatorname{Re} b)^2(1+\alpha^2) + \lambda^{-2}}{(n \operatorname{Re} b + \lambda^{-1})^2} = \frac{1}{n}$$

are given by

$$\lambda_{1,2} = \frac{1 \pm \sqrt{1 + (1-n)\alpha^2}}{n\alpha^2 \operatorname{Re} b}.$$

Therefore, if $\alpha < \sqrt{\frac{1}{n-1}}$, we have two real positive solutions. Now choose

$$\frac{1 - \sqrt{1 + (1-n)\alpha^2}}{n\alpha^2 \operatorname{Re} b} < \lambda < \frac{1 + \sqrt{1 + (1-n)\alpha^2}}{n\alpha^2 \operatorname{Re} b}$$

and

$$0 < \varepsilon \leq \frac{(n \operatorname{Re} b + \lambda^{-1})^2}{n(\operatorname{Re} b)^2(1+\alpha^2) + \lambda^{-2}} - n = \frac{2n\lambda^{-1}\operatorname{Re} b + (1-n)\lambda^{-2} - \alpha^2 n^2(\operatorname{Re} b)^2}{n(1+\alpha^2)(\operatorname{Re} b)^2 + \lambda^{-2}}.$$

Then $b\Delta$ satisfies the modified Cordes condition with λ and ε. $\qquad\square$

Lemma 4.3.2. *If $\mathcal{A}$ satisfies the modified parabolic Cordes condition (4.11) for some λ_0 and $\varepsilon > 0$ and $\sum_{i=1}^n \operatorname{Re} a_{ii}(x,t) > 0$ almost everywhere, then*

$$\frac{\sum_{i,j=1}^n |a_{ij}(x,t)|^2}{(\sum_{i=1}^n \operatorname{Re} a_{ii}(x,t))^2} \le \frac{1}{n-1+\varepsilon} \quad a.e. \ x \in \Omega, \ t \in J,$$

i.e. $\mathcal{A}$ satisfies the elliptic Cordes condition (4.2) for the same parameter $\varepsilon > 0$.

Proof. Consider the function $f : \mathbb{R}_+ \to \mathbb{R}_+$ defined by

$$f(\lambda) := \frac{(b + \lambda^{-1})^2}{a + \lambda^{-2}},$$

where $a, b > 0$. Then

$$f'(\lambda) = \frac{2\lambda^{-3}(b + \lambda^{-1})}{(a + \lambda^{-2})^2}(b - \lambda a),$$

so f reaches its sole maximum at $\lambda = \frac{b}{a}$. Fix $(x,t) \in Q$ and let

$$b = \sum_{i=1}^n \operatorname{Re} a_{ii}(x,t) \quad \text{and} \quad a = \sum_{i,j=1}^n |a_{ij}(x,t)|^2.$$

Then f takes its maximum at

$$\lambda_{(x,t)} = \frac{\sum_{i=1}^n \operatorname{Re} a_{ii}(x,t)}{\sum_{i,j=1}^n |a_{ij}(x,t)|^2}$$

and for almost every $(x,t) \in Q$ we have the estimate

$$\begin{aligned}
n + \varepsilon &\le \frac{(\sum_{i=1}^n \operatorname{Re} a_{ii}(x,t) + \lambda_0^{-1})^2}{\sum_{i,j=1}^n |a_{ij}(x,t)|^2 + \lambda_0^{-2}} \\
&\le \frac{(\sum_{i=1}^n \operatorname{Re} a_{ii}(x,t) + \lambda_{(x,t)}^{-1})^2}{\sum_{i,j=1}^n |a_{ij}(x,t)|^2 + \lambda_{(x,t)}^{-2}} \\
&= \frac{(\lambda_{(x,t)} \sum_{i=1}^n \operatorname{Re} a_{ii}(x,t) + 1)^2}{\lambda_{(x,t)}^2 \sum_{i,j=1}^n |a_{ij}(x,t)|^2 + 1} \\
&= \left(\frac{(\sum_{i=1}^n \operatorname{Re} a_{ii}(x,t))^2}{\sum_{i,j=1}^n |a_{ij}(x,t)|^2} + 1 \right)^2 \left(\frac{(\sum_{i=1}^n \operatorname{Re} a_{ii}(x,t))^2}{\sum_{i,j=1}^n |a_{ij}(x,t)|^2} + 1 \right)^{-1} \\
&= \frac{(\sum_{i=1}^n \operatorname{Re} a_{ii}(x,t))^2}{\sum_{i,j=1}^n |a_{ij}(x,t)|^2} + 1
\end{aligned}$$

which is what we wanted to show. □

We now introduce the spaces in which we want to solve the parabolic problem $u'(t) - A(t)u = f$. Fix $\lambda > 0$ so that $\mathcal{A}$ satisfies (4.10) and (4.11) with λ. Let

$$
\begin{aligned}
Y_p &:= L^p(J, W^{2,p}(\Omega) \cap W_0^{1,p}(\Omega)) \cap W^{1,p}(J, L^p(\Omega)), \\
Z_p &:= \{u \in Y_p : u(x,0) = 0 \text{ in } \Omega\},
\end{aligned}
$$

and for $u \in Y_p$, define

$$
\|u\|_\lambda := \left(\int_Q \left(\sum_{i,j=1}^n |D_{ij}u|^2 + \lambda^2|u_t|^2 \right)^{p/2} \right)^{1/p}.
$$

We already saw in Theorem 3.3.2 that for $1 < p \leq 2$, the Laplacian with domain $D(\Delta) = W^{2,p}(\Omega) \cap W_0^{1,p}(\Omega)$ has maximal L^q-regularity. Therefore, given $f \in L^p(Q)$, $1 < p \leq 2$, there exists a unique $u \in Z_p$ with $u_t - \Delta u = f$. Rescaling the time, by setting $v(t,x) = u(\lambda^{-1}t, x)$, we see that for every $\lambda > 0$ and $f \in L^p(Q)$ there exists a unique solution to $\lambda v_t - \Delta v = f$ in Z_p. This observation leads us to the following proposition.

Proposition 4.3.3. *For $1 < p \leq 2$, $(Z_p, \|\cdot\|_\lambda)$ is a Banach space.*

Proof. We need to show that in Z_p the norm $\|\cdot\|_\lambda$ is equivalent to the usual norm given by

$$
\|u\| = \left(\int_Q \left(\sum_{|\alpha| \leq 2} |D^\alpha u|^2 + |u_t|^2 \right)^{p/2} \right)^{1/p}.
$$

Let $u \in Z_p$. Obviously,

$$
\|u\|_\lambda \leq C \left(\|u\|_{L^p(J, W^{2,p}(\Omega))} + \|u\|_{W^{1,p}(J, L^p(\Omega))} \right).
$$

Now let $f = (\lambda\partial_t - \Delta)u$. Then by Theorem 3.3.2 and the observation preceding this proposition, we have

$$
\begin{aligned}
\|u\|_{L^p(J, W^{2,p}(\Omega))} + \|u\|_{W^{1,p}(J, L^p(\Omega))} &\leq C\|f\|_{L^p(Q)} \\
&= C\|(\lambda\partial_t - \Delta)u\|_{L^p(Q)}
\end{aligned}
$$

$$\leq \ C \left\Vert u\right\Vert_\lambda. \qquad\qquad \square$$

Consider the following non-autonomous parabolic problem. For $t \in J$ denote by $A(t)$ the operator defined by

$$(4.12) \qquad\qquad A(t)u \ := \ \mathcal{A}(t)u,$$

$$(4.13) \qquad\qquad D(A(t)) \ := \ Z_p.$$

Given $f \in L^p(Q)$, find $u \in Z_p$ such that

$$(4.14) \qquad\qquad u'(t) - A(t)u = f \text{ in } Q.$$

Remarks 4.3.4. From Theorem 3.3.2, we know that the operator $\partial_t - \Delta$ with domain Z_p is a closed operator in $L^p(Q)$, $1 < p \leq 2$. We will show that this is also the case for $\partial_t - A(t)$ when $\mathcal{A}$ satisfies the parabolic Cordes condition and p is close to 2 (Theorem 4.3.9). Note that even though the operator depends on t, we choose the domain independent of t.

4.3.1   The parabolic case for $p = 2$

Our result for $p = 2$ is the following:

Proposition 4.3.5. *Let Ω be a bounded convex domain in $\mathbb{R}^n$. Let $A(t)$ be an operator defined as in (4.12) and (4.13) and assume that (4.10) and (4.11) are satisfied. Then, given $f \in L^2(Q)$, there is a unique $u \in Z_2$ solving (4.14).*

Before proving the proposition, we show two useful lemmas.

Lemma 4.3.6. *For $u \in Z_2$, we have*

$$\mathrm{Re} \ \int_Q \Delta u \ \overline{u}_t \leq 0.$$

Proof.   First, let $J = [0, T]$. Then, by a density argument (cf. [Maz85, Theorem 1.1.6/2]), it is enough to consider $u \in C^\infty(\overline{Q})$ with $u(x, 0) = 0$ and $u|_{\partial\Omega} = 0$. For such a function u we have

$$\frac{\mathrm{d}}{\mathrm{d}t}|\nabla u|^2 \ = \ \nabla u \frac{\mathrm{d}}{\mathrm{d}t}\nabla\overline{u} + \nabla\overline{u}\frac{\mathrm{d}}{\mathrm{d}t}\nabla u$$

$$= 2 \operatorname{Re} \nabla u \nabla \bar{u}_t.$$

Therefore,

$$
\begin{aligned}
\operatorname{Re} \int_Q \Delta u \, \bar{u}_t &= -\operatorname{Re} \int_Q \nabla u \nabla \bar{u}_t \\
&= -\frac{1}{2} \int_Q \frac{\mathrm{d}}{\mathrm{d}t} |\nabla u|^2 \\
&= -\frac{1}{2} \int_\Omega |\nabla u(x,T)|^2 \leq 0.
\end{aligned}
$$

If $J = [0,\infty)$, let $Q_T = \Omega \times [0,T]$. Then, by the first case, we get

$$
\operatorname{Re} \int_Q \Delta u \, \bar{u}_t = \lim_{T \to \infty} \operatorname{Re} \int_{Q_T} \Delta u \, \bar{u}_t \leq 0.
$$

$\square$

Lemma 4.3.7. *Let $f \in L^2(Q)$, $\lambda > 0$ and suppose that u satisfies $\lambda u_t - \Delta u = f$. Then $\|u\|_\lambda \leq \|f\|_{L^2}$.*

Proof. Using Proposition 4.2.5 and Lemma 4.3.6, we obtain

$$
\begin{aligned}
\|u\|_\lambda^2 &= \int_Q \left(\sum_{i,j=1}^n |D_{ij}u|^2 + \lambda^2 |u_t|^2 \right) \\
&\leq \int_Q \left(|\Delta u|^2 + \lambda^2 |u_t|^2 \right) \\
&\leq \int_Q \left(|\Delta u|^2 + \lambda^2 |u_t|^2 - 2\lambda \operatorname{Re} \Delta u \, \bar{u}_t \right) \\
&= \int_Q |\Delta u - \lambda u_t|^2 \\
&= \|f\|_{L^2}^2 .
\end{aligned}
$$

For the last but one equality, note that for $z_1, z_2 \in \mathbb{C}$ we have

$$|z_1 - z_2|^2 = |z_1|^2 + |z_2|^2 - 2 \operatorname{Re} z_1 \operatorname{Re} z_2 - 2 \operatorname{Im} z_1 \operatorname{Im} z_2$$

and

$$\operatorname{Re} z_1 \operatorname{Re} z_2 + \operatorname{Im} z_1 \operatorname{Im} z_2 = \operatorname{Re} z_1 \bar{z}_2.$$

$\square$

We are now able to prove the Proposition 4.3.5.

Proof.   We rewrite problem (4.14) as in the elliptic case. It is equivalent to finding $u \in Z_2$ such that

$$\lambda u_t - \Delta u = \alpha f + \sum_{i,j=1}^{n} (\alpha a_{ij} - \delta_{ij})D_{ij}u + (\lambda - \alpha)u_t$$

where we choose

$$\alpha(x,t) = \frac{\sum_{i=1}^{n} \operatorname{Re} a_{ii}(x,t) + \lambda^{-1}}{\sum_{i,j=1}^{n} |a_{ij}(x,t)|^2 + \lambda^{-2}}.$$

Note that since $|\alpha(x,t)| \leq \lambda^2(\sum_{i=1}^{n} \|a_{ii}\|_\infty + \lambda^{-1})$, the function α is bounded. Now, using Theorem 3.3.2, we define a map $\Phi : Z_2 \to Z_2$ by $\Phi w := u$ where, for given $w \in Z_2$, u is the unique solution in Z_2 to

$$\lambda u_t - \Delta u = \alpha f + \sum_{i,j=1}^{n} (\alpha a_{ij} - \delta_{ij})D_{ij}w + (\lambda - \alpha)w_t.$$

Let $w, \overline{w} \in Z_2$ and $u = \Phi(w), \overline{u} = \Phi(\overline{w})$. Then, by Lemma 4.3.7 and the Cauchy-Schwarz Inequality,

$$
\begin{aligned}
\|\Phi w - \Phi \overline{w}\|_\lambda^2 &= \|u - \overline{u}\|_\lambda^2 \\
&\leq \left\| \sum_{i,j=1}^{n} (\alpha a_{ij} - \delta_{ij})D_{ij}(w - \overline{w}) + (\lambda - \alpha)(w_t - \overline{w}_t) \right\|_{L^2}^2 \\
&\leq \int_Q \left(\sum_{i,j=1}^{n} |\alpha a_{ij} - \delta_{ij}|^2 + \frac{(\lambda - \alpha)^2}{\lambda^2} \right) \\
&\qquad \cdot \left(\sum_{i,j=1}^{n} |D_{ij}(w - \overline{w})|^2 + \lambda^2 |w_t - \overline{w}_t|^2 \right) \\
&\leq (1 - \varepsilon) \|w - \overline{w}\|_\lambda^2,
\end{aligned}
$$

as

$$
\begin{aligned}
\sum_{i,j=1}^{n} |\alpha a_{ij} - \delta_{ij}|^2 &+ \frac{(\lambda - \alpha)^2}{\lambda^2} \\
&= \alpha^2 \sum_{i,j=1}^{n} |a_{ij}|^2 - 2\alpha \sum_{i=1}^{n} \operatorname{Re} a_{ii} + n + 1 - \frac{2\alpha}{\lambda} + \frac{\alpha^2}{\lambda^2}
\end{aligned}
$$

$$= -\frac{(\sum_{i=1}^{n} \operatorname{Re} a_{ii} + \lambda^{-1})^2}{\sum_{i,j=1}^{n} |a_{ij}|^2 + \lambda^{-2}} + n + 1$$

$$\leq 1 - \varepsilon.$$

Applying the Banach Fixed Point Theorem to Φ ends the proof.   □

4.3.2   The parabolic case for $p \neq 2$

Just as in the elliptic case, the first step in examining the parabolic case for $p \neq 2$ is to transfer Lemma 4.3.7 to $L^p(Q)$.

Proposition 4.3.8. *Let $1 < p \leq 2$ and $\lambda > 0$. Then, given $f \in L^p(Q)$, there exists a constant C_p and a unique $u \in Z_p$ satisfying $(\lambda \partial_t - \Delta)u = f$ with $\|u\|_\lambda \leq C_p \|f\|_{L^p}$. Furthermore, for any $\varepsilon \in (0,1)$, there exists $\delta > 0$ such that for $p \in (2-\delta, 2]$, the constant C_p satisfies the estimate $C_p < (1-\varepsilon)^{-1/2}$.*

Proof.   Using the maximal regularity of the Laplacian with domain $D(\Delta) = W^{2,p}(\Omega) \cap W_0^{1,p}(\Omega)$ (see Theorem 3.3.2), we can define the operator T on $L^p(Q)$ by

$$Tf := \left(\sum_{i,j=1}^{n} |D_{ij}u|^2 + \lambda^2 |u_t|^2 \right)^{1/2},$$

where u is the unique solution of $(\lambda \partial_t - \Delta)u = f$ in Z_p. Then $\|Tf\|_p \leq C_p \|f\|_p$. Moreover, by Lemma 4.3.7, $C_2 = 1$. Thus, by the Riesz-Thorin Interpolation Theorem, C_p is close to 1 for p near 2.   □

A similar calculation as in the proof of Proposition 4.3.5 then proves the following.

Theorem 4.3.9. *Let Ω be a bounded convex domain in $\mathbb{R}^n$. For $t \in J$, let $A(t)$ be an operator defined as in (4.12) and (4.13) and satisfying (4.10) and (4.11) for some $\varepsilon \in (0,1)$. Then there exists $\delta > 0$ such that for $p \in (2-\delta, 2]$ and $f \in L^p(Q)$, there is a unique $u \in Z_p$ satisfying $(\partial_t - A(t))u = f$ in Q.*

This gives us maximal regularity and the closed graph theorem then yields the following a priori estimate:

Corollary 4.3.10. *Under the assumptions of Theorem 4.3.9, A has maximal L^p-regularity on $L^p(\Omega)$ and the solution to (4.14) satisfies*

$$\|u\|_{L^p(Q)} + \|u'\|_{L^p(Q)} + \|Au\|_{L^p(Q)} \leq C \|f\|_{L^p(Q)} .$$

In the autonomous case, i.e. when the coefficients of $\mathcal{A}$ are independent of t, we also obtain that A is the generator of a semigroup.

Corollary 4.3.11. *Assume that the coefficients a_{ij} of $\mathcal{A}$ do not depend on t. Then, under the assumptions of Theorem 4.3.9, A generates an analytic semigroup T on $L^p(\Omega)$ for $p \in (2-\delta, 2]$. Moreover, if $J = [0, \infty)$, the growth bound $\omega(T)$ of the semigroup T is negative.*

Proof.   It is well known that maximal regularity for an operator implies that the operator generates an analytic semigroup (cf. Section 1.3 and the references given there). That the growth bound of the generated semigroup is negative in the case that $J = [0, \infty)$ is proved in [HP97].   □

In the autonomous case, we also get L^q-L^p-estimates.

Corollary 4.3.12. *Assume that the coefficients a_{ij} of $\mathcal{A}$ do not depend on t. If A and p satisfy the assumptions of Theorem 4.3.9, then for every $f \in L^q(J, L^p(\Omega))$, $1 < q < \infty$, the equation*

$$\begin{cases} u' - \mathcal{A}u = f & in \ Q, \\ \quad\quad\ u = 0 & on \ \partial\Omega, \\ \ \ u(\cdot, 0) = 0 & in \ \Omega \end{cases}$$

has a unique solution $u \in L^q(J, W^{2,p}(\Omega) \cap W_0^{1,p}(\Omega)) \cap W^{1,q}(J, L^p(\Omega))$ and

$$\|u\|_{L^q(J,L^p(\Omega))} + \|u'\|_{L^q(J,L^p(\Omega))} + \|Au\|_{L^q(J,L^p(\Omega))} \leq C \|f\|_{L^q(J,L^p(\Omega))} .$$

Proof.   This is due to Sobolevskii [Sob64] and Dore [Dor93, in particular Theorem 4.2]; see also [DHP03].   □

Returning to the non-autonomous case, we have the following abstract result on maximal regularity of the operators $A(t_0)$.

Proposition 4.3.13. *Let X be a Banach space, $\{B(t),\ t \in J\}$ a family of operators on X with domain $D(B(t)) = D$ for all $t \in J$. Assume that for every $f \in L^q(J, X)$, the equation*

$$(4.15) \qquad u'(t) - B(t)u(t) = f(t)$$

has a unique solution $u \in Y := L^q(J, D) \cap W^{1,q}(J, X)$. Furthermore, assume the map $\phi : J \to \mathcal{L}(Y, L^q(J, X))$, $\phi(t) = B(t)$ is continuous in some $t_0 \in J$. Then for every $f \in L^q(J, X)$, there is a unique solution in Y to

$$u'(t) - B(t_0)u(t) = f(t).$$

Proof. Let $M : L^q(J, X) \to Y$ be the solution operator to (4.15). Then by the closed graph theorem, there exists a constant C_M such that

$$\|Mf\|_Y \leq C_M \|f\|_{L^q(J,X)}.$$

By continuity of ϕ, there exists $\delta > 0$ such that for $|t - t_0| < \delta$,

$$\|B(t_0) - B(t)\|_{\mathcal{L}(Y, L^q(J,X))} \leq \frac{1}{2C_M}.$$

Then

$$\|(B(t_0) - B(t))M\|_{\mathcal{L}(L^q(J,X))} \leq \frac{1}{2}.$$

This implies the invertibility in $\mathcal{L}(Y, L^q(J, X))$ of the operator

$$\partial_t - B(t_0) = (1 - (B(t_0) - B(t))M)(\partial_t - B(t)),$$

which ends the proof. $\qquad\square$

In our situation, this result implies maximal regularity for $A(t_0)$ whenever we have maximal regularity for the non-autonomous problem and continuity in t of the coefficients a_{ij}. However, as for almost all t_0, $A(t_0)$ obviously satisfies the Cordes condition whenever $A(t)$ does, the continuity condition is not needed here and we get the following result:

Corollary 4.3.14. *Under the assumptions of Theorem 4.3.9, for almost every $t_0 \in J$, the operator $A(t_0)$ generates an analytic semigroup T_{t_0} on*

$L^p(\Omega)$ for $p \in (2 - \delta, 2]$. Furthermore, $A(t_0)$ has maximal L^p-regularity and the solution to

$$
\begin{cases}
u' - \mathcal{A}(t_0)u = f & in\ Q, \\
u = 0 & on\ \partial\Omega, \\
u(\cdot, 0) = 0 & in\ \Omega
\end{cases}
$$

satisfies

$$\|u\|_{L^p(Q)} + \|u'\|_{L^p(Q)} + \|\mathcal{A}(t_0)u\|_{L^p(Q)} \le C \|f\|_{L^p(Q)}.$$

The following corollary finally deals with the particular situation of two dimensions. Unfortunately, unlike in the case of elliptic equations, symmetry of the operator together with strong ellipticity is not sufficient to verify the modified parabolic Cordes condition. In order to be able to choose one fixed $\lambda > 0$ for almost every $(x,t) \in Q$, we also need a condition guaranteeing that the coefficients do not vary too much.

Corollary 4.3.15. *Let Ω be a bounded convex domain in $\mathbb{R}^2$. Let A be defined as in (4.12) and (4.13) and assume that $a_{ij} \in L^\infty(Q, \mathbb{R})$ such that $a_{12} = a_{21}$. Assume that A is strongly elliptic, i.e. there exists $\mu > 0$ such that*

$$\sum_{i,j=1}^{2} a_{ij}(x,t)\xi_i\xi_j \ge \mu|\xi|^2, a.e. (x,t) \in Q.$$

Furthermore, we assume that

$$
\begin{aligned}
(4.16)\ \operatorname*{esssup}_{(x,t)\in Q}\ &\left(a_{11}(x,t) + a_{22}(x,t) - 2\sqrt{a_{11}(x,t)a_{22}(x,t) - a_{12}(x,t)^2} \right) \\
&< \operatorname*{essinf}_{(x,t)\in Q}\ \left(a_{11}(x,t) + a_{22}(x,t) + 2\sqrt{a_{11}(x,t)a_{22}(x,t) - a_{12}(x,t)^2} \right).
\end{aligned}
$$

Then, there exists $\delta > 0$ such that for $p \in (2 - \delta, 2]$ and $f \in L^p(Q)$, there is a unique $u \in Z_p$ satisfying $(\partial_t - A)u = f$ in Q.

Proof. By setting $\xi_1 = 1, \xi_2 = 0$ and $\xi_1 = 0, \xi_2 = 1$, we see that $a_{11}(x,t) \ge \mu$ and $a_{22}(x,t) \ge \mu$, so (4.10) is satisfied for any $\lambda > 0$. It remains to show that (4.11) holds. For this, we need to show that there exists $\lambda > 0$ such that for some $\varepsilon > 0$

$$(2+\varepsilon)(a_{11}^2 + 2a_{12}^2 + a_{22}^2 + \lambda^2) \le a_{11}^2 + 2a_{11}a_{22} + a_{22}^2 + 2\lambda(a_{11} + a_{22}) + \lambda^2.$$

Due to the boundedness of the coefficients, this is equivalent to finding $\lambda > 0$ such that

$$2(a_{11}^2 + 2a_{12}^2 + a_{22}^2 + \lambda^2) < a_{11}^2 + 2a_{11}a_{22} + a_{22}^2 + 2\lambda(a_{11} + a_{22}) + \lambda^2,$$

in other words,

$$\lambda^2 - 2\lambda(a_{11} + a_{22}) + a_{11}^2 + 4a_{12}^2 + a_{22}^2 - 2a_{11}a_{22} < 0.$$

We consider the equation

$$\lambda^2 - 2\lambda(a_{11} + a_{22}) + a_{11}^2 + 4a_{12}^2 + a_{22}^2 - 2a_{11}a_{22} = 0.$$

Its roots are given by

$$\lambda_\pm = a_{11}(x,t) + a_{22}(x,t) \pm 2\sqrt{a_{11}(x,t)a_{22}(x,t) - a_{12}(x,t)^2}.$$

Now, from the ellipticity condition with $\xi_1 = \sqrt{a_{22}}$ and $\xi_2 = \pm\sqrt{a_{11}}$, we get

$$(4.17) \qquad 2a_{11}a_{22} \pm 2a_{12}\sqrt{a_{11}a_{22}} \geq \mu(a_{11} + a_{22}) > 0.$$

This yields $a_{12}^2 < a_{11}a_{22}$ and therefore $\lambda_\pm \in \mathbb{R}$. Also, λ_+ is positive, as $\lambda_+ > 2\mu$. However, we still have the problem that $\lambda_\pm$ depends on $(x,t) \in Q$. The condition (4.16) now guarantees that we can choose $\lambda > 0$ independent of $(x,t) \in Q$ such that $\lambda_-(x,t) < \lambda < \lambda_+(x,t)$. Having done this, we can choose $\varepsilon > 0$ so that (4.11) is satisfied. The statement then follows from Theorem 4.3.9. $\qquad\square$

Remarks 4.3.16. 1. The condition (4.16) reflects the fact that in the modified Cordes condition (4.11), λ needs to be chosen according to the size of the coefficients $a_{ij}(x,t)$. If the coefficients vary too much over time or space, it may not be possible to find one fixed λ to "suit all sizes". This however is needed to be able to solve the equation $\lambda u' - \Delta u = f$ in Z_p.

2. Obviously for constant coefficients a_{ij}, (4.16) is always satisfied for strongly elliptic operators.

3. If A is strongly elliptic with ellipticity constant μ and

$$M := \max\{\|a_{11}\|_\infty, \|a_{22}\|_\infty\} < 2\mu,$$

then (4.16) is satisfied.

Proof.   From (4.17), we get the estimate

$$a_{11}a_{22} \pm a_{12}\sqrt{a_{11}a_{22}} \geq \mu^2.$$

Rearranging the terms and squaring gives

$$a_{11}a_{22} - 2\mu^2 + \frac{\mu^4}{a_{11}a_{22}} \geq a_{12}^2,$$

and therefore

$$a_{11}a_{22} - a_{12}^2 \geq \frac{2\mu^2 a_{11}a_{22} - \mu^4}{a_{11}a_{22}} \geq \frac{\mu^4}{M^2}.$$

We finally have the following estimates

$$\lambda_- \;\leq\; 2M - 2\frac{\mu^2}{M},$$

$$\lambda_+ \;\geq\; 2\mu + 2\frac{\mu^2}{M}.$$

Then the condition $M < 2\mu$ guarantees $\lambda_- < 3\mu < \lambda_+$.   □

Chapter 5

The Ornstein-Uhlenbeck operator in Lipschitz domains

We want to consider the following problem. Let M be a real-valued constant coefficient $n \times n$-matrix. Let K be a compact set in $\mathbb{R}^n$ with Lipschitz boundary, and let Ω be the exterior domain $\Omega = \mathbb{R}^n \setminus K$. On this domain we consider the following equation.

$$
(5.1) \quad
\begin{cases}
\partial_t u(x,t) = \Delta u(x,t) + Mx \cdot \nabla u(x,t) & \text{in } \Omega \times \mathbb{R}_+, \\
u(x,t) = 0 & \text{on } \partial\Omega \times \mathbb{R}_+, \\
u(x,0) = u_0(x) & \text{for } x \in \Omega.
\end{cases}
$$

Operators of the form $\Delta + Mx \cdot \nabla$ are called *Ornstein-Uhlenbeck operators*.[1] One setting where Ornstein-Uhlenbeck operators arise naturally is when transforming equations such as the heat equation or the Navier-Stokes equations from rotating domains to a fixed domain Ω (see [His99a]). In bounded domains, Ornstein-Uhlenbeck operators can be considered as relatively bounded perturbations of the Laplacian. If the domain is not bounded however, the term Mx is unbounded which makes the problem much more difficult to consider.

In order to examine Ornstein-Uhlenbeck operators in exterior domains, we use a cut-off procedure previously used by Hishida ([His99a], [His99b] and

---

[1] Note that x is actually the space variable, so the notation $\Delta + Mx \cdot \nabla$ is formally bad notation, however it is the standard notation and we will use it in the following.

[His00]) and study the corresponding resolvent problem first in the whole space $\mathbb{R}^n$ and then on a bounded Lipschitz domain D. Combining these two solutions appropriately will solve the resolvent problem on the domain Ω and via the Lumer-Phillips Theorem we will obtain that the operator $\Delta + Mx \cdot \nabla$ with suitable domain generates a C_0-semigroup on $L^p(\Omega)$ for a p-interval around 2. In domains satisfying a uniform outer ball condition, we prove that for $1 < p \leq 2$, the domain of the Ornstein-Uhlenbeck operator on $L^p(\Omega)$ is contained in $W^{2,p}(\Omega)$ and show L^p-L^q-smoothing estimates.

5.1   Ornstein-Uhlenbeck operators on $L^p(\mathbb{R}^n)$

On the whole space $\mathbb{R}^n$, Ornstein-Uhlenbeck operators have been extensively studied by many authors. Here, we refer to the papers by Da Prato, Lunardi and Vespri, [DPL95] and [LV96], and by Metafune, Prüss, Rhandi and Schnaubelt (see [Met01], [MPRS02], [MPRS04] and [PRS04]).

We define the following operator. For $1 < p < \infty$, let

(5.2)
$$
\begin{cases}
A_{\mathbb{R}^n} u(x) := \Delta u(x) + Mx \cdot \nabla u(x), \ x \in \mathbb{R}^n, \\
D(A_{\mathbb{R}^n}) := \{ u \in W^{2,p}(\mathbb{R}^n) : Mx \cdot \nabla u \in L^p(\mathbb{R}^n) \}.
\end{cases}
$$

We gather the main results on the Ornstein-Uhlenbeck operator $A_{\mathbb{R}^n}$ in the next theorem.

Theorem 5.1.1. *Let $1 < p < \infty$ then the operator $A_{\mathbb{R}^n}$ as defined in (5.2) generates a positive C_0-semigroup on $L^p(\mathbb{R}^n)$ and the semigroup $(e^{tA_{\mathbb{R}^n}})_{t \geq 0}$ has the explicit representation*

$$
e^{tA_{\mathbb{R}^n}} f(x) = \frac{1}{(4\pi)^{n/2}(\det Q_t)^{1/2}} \int_{\mathbb{R}^n} f(e^{tM}x - y) e^{-\frac{1}{4}(Q_t^{-1}y, y)} dy, \ x \in \mathbb{R}^n, \ t > 0,
$$

with $Q_t := \int_0^t e^{sM} e^{sM^T} \, ds$ for $t > 0$.

Moreover, there exists $\lambda_0 \in \mathbb{R}$ such that for $\lambda > \lambda_0$ the unique solution of the resolvent problem

$$
(\lambda - A_{\mathbb{R}^n}) u = f
$$

satisfies the estimate

$$(5.3) \qquad \|u\|_{W^{2,p}(\mathbb{R}^n)} + \|Mx \cdot \nabla u\|_{L^p(\mathbb{R}^n)} \le C_\lambda \|f\|_{L^p(\mathbb{R}^n)},$$

for some constant C_λ depending on λ, and there exist constants C, ω independent of λ such that the solution to the resolvent problem satisfies the gradient estimate

$$(5.4) \qquad \|\nabla u\|_{L^p(\mathbb{R}^n)} \le \frac{C}{|\lambda - \omega|^{\frac{1}{2}}} \|f\|_{L^p(\mathbb{R}^n)}.$$

Furthermore, if $1 < p \le q \le \infty$, then there exist constants C, ω such that

$$(5.5) \qquad \left\|e^{tA_{\mathbb{R}^n}} f\right\|_q \le C t^{-\frac{n}{2}(\frac{1}{p} - \frac{1}{q})} e^{\omega t} \|f\|_p, \; t > 0, \; f \in L^p(\mathbb{R}^n),$$

and for the gradient we have the estimate

$$(5.6) \qquad \left\|\nabla e^{tA_{\mathbb{R}^n}} f\right\|_q \le C t^{-\frac{1}{2} - \frac{n}{2}(\frac{1}{p} - \frac{1}{q})} e^{\omega t} \|f\|_p, \; t > 0, \; f \in L^p(\mathbb{R}^n).$$

Proof. By [PRS04, Theorem 2.4], we obtain that $A_{\mathbb{R}^n}$ generates a positive C_0-semigroup on $L^p(\mathbb{R}^n)$ and the explicit representation can be found in [Met01]. The estimate (5.3) then follows from the closed graph theorem. For the gradient estimate (5.4), using [HS04, Proposition 3.4], we see that the semigroup generated by $A_{\mathbb{R}^n}$ satisfies the estimate

$$(5.7) \qquad \left\|\nabla e^{tA_{\mathbb{R}^n}} f\right\|_p \le \frac{C}{t^{\frac{1}{2}}} e^{\omega t} \|f\|_p$$

for some $\omega \in \mathbb{R}$. For $\lambda > \lambda_0$, the resolvent is given by the Laplace transform of the semigroup (cf. Section 1.3) and we get

$$\begin{aligned}
\|\nabla R(\lambda, A_{\mathbb{R}^n}) f\|_p &= \left\| \int_0^\infty e^{-\lambda t} \nabla e^{tA_{\mathbb{R}^n}} f \, \mathrm{d}t \right\|_p \\
&\le C \int_0^\infty t^{-\frac{1}{2}} e^{(\omega - \lambda)t} \, \mathrm{d}t \, \|f\|_p \\
&\le C \frac{\Gamma(\frac{1}{2})}{(\lambda - \omega)^{\frac{1}{2}}} \|f\|_p.
\end{aligned}$$

As all integrals exist and the operator ∇ with $D(\nabla) = W^{1,p}(\mathbb{R}^n)$ is closed, the differentiation under the integral is justified. Finally, (5.5) and (5.6) follow from [HS04, Proposition 3.4 and Lemma 3.5]. $\qquad\square$

5.2 The drift operator

We have already studied the Laplacian in Lipschitz domains in the Chapters 2 and 3. We now take a closer look at the drift term before combining all our results to obtain results for the Ornstein-Uhlenbeck operator. We first introduce the drift operator B_Ω on a domain $\Omega \subseteq \mathbb{R}^n$:

$$(5.8) \qquad \begin{cases} B_\Omega u(x) := Mx \cdot \nabla u(x), \ x \in \Omega, \\ D(B_\Omega) := \{u \in W^{1,p}(\Omega) : Mx \cdot \nabla u \in L^p(\Omega)\}. \end{cases}$$

Note that the condition $Mx \cdot \nabla u \in L^p(\Omega)$ is trivially satisfied in bounded domains for $u \in W^{1,p}(\Omega)$. Moreover, in this case the drift term is a relatively bounded perturbation of the Laplacian:

Proposition 5.2.1. *Let D be a bounded Lipschitz domain. Then the drift operator B_D on D is relatively bounded by the weak Dirichlet-Laplacian $\Delta_{p,w}^D$ (cf. Definition 3.1.1) in $L^p(D)$ for $(3+\varepsilon)' < p < 3+\varepsilon$ where $\varepsilon > 0$ depends only on the Lipschitz constant of D. The relative bound is given by zero.*

Proof. Obviously we have that $D(B_D) \supseteq D(\Delta_{p,w}^D)$. Let $u \in D(\Delta_{p,w}^D)$. By Theorem 2.1.1, we can find $\delta > 0$ such that

$$\|u\|_{1+\delta,p} \le C \|\Delta u\|_p \quad \text{for} \quad (3+\varepsilon)' < p < 3+\varepsilon.$$

As the weak Dirichlet-Laplacian generates a contractive C_0-semigroup by Theorem 3.3.1, we have the resolvent estimate

$$\|u\|_p \le \frac{C}{\lambda} \|(\lambda - \Delta)u\|_p \quad \text{for} \quad \lambda > 0, \quad (3+\varepsilon)' < p < 3+\varepsilon.$$

Set $\Theta = \delta/(1+\delta) \in (0,1)$. Then, using the complex interpolation method (cf. [BL76]) and Jensen's inequality, we obtain the estimate

$$\begin{aligned} \|u\|_{1,p} &\le \left(C \|\Delta u\|_p\right)^{1-\Theta} \left(\frac{C}{\lambda} \|(\lambda - \Delta)u\|_p\right)^{\Theta} \\ &= C\lambda^{-\Theta} \|\Delta u\|_p^{1-\Theta} \|(\lambda - \Delta)u\|_p^{\Theta} \\ &\le C\lambda^{-\Theta}((1-\Theta) \|\Delta u\|_p + \Theta \|(\lambda - \Delta)u\|_p) \\ &= C\lambda^{-\Theta} \|\Delta u\|_p + C\lambda^{1-\Theta} \|u\|_p \end{aligned}$$

As we can choose $\lambda > 0$ arbitrarily large, this concludes the proof. $\square$

The next step is to prove dissipativity for the drift operator. To do this, we need to integrate by parts. We first prove that this is possible in our situation.

Proposition 5.2.2. *Let Ω be a Lipschitz domain in $\mathbb{R}^n$, $u \in W_0^{1,p}(\Omega)$, $\phi \in C_c^\infty(\mathbb{R}^n)$ and $1 < p < \infty$. Then we have*

$$
\begin{aligned}
(5.9) \qquad \int_\Omega \phi\, \overline{u}|u|^{p-2}\partial_i u \;=\; &- \int_\Omega \phi\, |u|^{p-2}u\partial_i\overline{u} \\
&- \int_\Omega \phi\, \frac{p-2}{2}|u|^{p-2}(\overline{u}\partial_i u + u\partial_i\overline{u}) \\
&- \int_\Omega (\partial_i\phi)\, |u|^p.
\end{aligned}
$$

Proof. The proof is based on the proof of Theorem 3.1 in [MS05]. As we only want to integrate by parts once, it is much simpler. For $p \geq 2$, we have $\phi\, \overline{u}|u|^{p-2} \in W^{1,p'}(\Omega)$ and we can directly integrate by parts. (5.9) then follows from a simple calculation. However, for $p < 2$ this is no longer the case and we divide the proof into six short steps:

Step 1: Actual integration by parts. For $u \in C_c^\infty(\Omega)$ and any $\delta > 0$, we have the equality

$$
\begin{aligned}
\int_\Omega \phi\, \overline{u}(|u|^2 + \delta)^{(p-2)/2}\partial_i u \;=\; &- \int_\Omega \phi\, (|u|^2 + \delta)^{(p-2)/2}u\partial_i\overline{u} \\
(5.10) \qquad &- \frac{p-2}{2}\int_\Omega \phi\, (|u|^2 + \delta)^{(p-4)/2}|u|^2(\overline{u}\partial_i u + u\partial_i\overline{u}) \\
&- \int_\Omega (\partial_i\phi)\, (|u|^2 + \delta)^{(p-2)/2}|u|^2.
\end{aligned}
$$

This follows from integrating by parts and the same calculation used in the case $p \geq 2$.

Step 2: (5.10) holds for $u \in W_0^{1,p} \cap L^\infty(\Omega)$. Approximate u in $W^{1,p}(\Omega)$ by $C_c^\infty(\Omega)$-functions u_n. Then (5.10) holds for every u_n. We may assume, possibly by passing to a subsequence, that $u_n \to u$ and $\nabla u_n \to \nabla u$ almost everywhere (see e.g. [Els96, Korollar VI.2.7]) and that $\|u_n\|_\infty \leq C \|u\|_\infty$. Then, recalling that $p' > p$, we have the following estimate

$$
\int_\Omega |u_n - u|^{p'} \;\leq\; \int_\Omega |u_n - u|^p \|u_n - u\|_\infty^{p'-p}
$$

$$\leq \ ((C+1)\,\|u\|_\infty)^{p'-p}\,\|u_n - u\|_p^p \to 0,$$

i.e. $u_n \to u$ also in $L^{p'}(\Omega)$. Since all of the terms ϕ, $(|u_n|^2 + \delta)^{(p-4)/2}$, $(|u_n|^2 + \delta)^{(p-2)/2}$ and $|u_n|^2$ are uniformly bounded in $L^\infty(\Omega)$, by Lebesgue's Dominated Convergence Theorem, we can take the limit into all four of the integrals. Pointwise convergence a.e. then implies that (5.10) holds for $u \in W_0^{1,p} \cap L^\infty(\Omega)$.

Step 3: The limit $\delta \to 0$. Let $u \in W_0^{1,p} \cap L^\infty(\Omega)$. Since, for all $\alpha < 0$, terms of the form $(|u|^2 + \delta)^\alpha$ can be estimated by $|u|^{2\alpha}$, and all integrals appearing in (5.9) exist for u, using the pointwise convergence and by dominated convergence, (5.9) holds for u.

Step 4: A characterisation of $W^{1,p}(\mathbb{R}^n)$. For $u \in L^p(\mathbb{R}^n)$ the following are equivalent:

1. $u \in W^{1,p}(\mathbb{R}^n)$

2. for every $i = 1, ..., n$ there exists an $(n-1)$-dimensional null set $N_i \subseteq \mathbb{R}^{n-1}$ such that for any $\tilde{x}_i := (x_1, ..., x_{i-1}, x_{i+1}, ..., x_n) \notin N_i$ the function $u(\tilde{x}_i, \cdot) \in W^{1,p}(\mathbb{R})$ and its derivative coincides a.e. with $\partial_i u$ in $L^p(\mathbb{R}^n)$.

This basically follows from an application of Fubini's Theorem. See [Zie89, Theorem 2.1.4 and Remark 2.1.5] for more details.

Step 5: (5.9) holds for $u \in W^{1,p}(\mathbb{R}^n)$. Let $u \in W^{1,p}(\mathbb{R}^n)$. Then by the fourth step, $u(\tilde{x}_i, \cdot) \in W^{1,p}(\mathbb{R})$. By Sobolev embeddings, $W^{1,p}(\mathbb{R}) \subseteq L^\infty(\mathbb{R})$, so, by Step 3, (5.9) holds for $u(\tilde{x}_i, \cdot)$ with $\Omega = \mathbb{R}$. Integrating both sides over the remaining $n-1$ variables proves the claim.

Step 6: Proof of the proposition. Let $u \in W_0^{1,p}(\Omega)$ and denote the trivial extension of u to $\mathbb{R}^n$ by $\tilde{u}$. Then $\tilde{u} \in W^{1,p}(\mathbb{R}^n)$, so we have

$$(5.11) \qquad \int_{\mathbb{R}^n} \phi\, \overline{\tilde{u}} |\tilde{u}|^{p-2} \partial_i \tilde{u} \ = \ -\int_{\mathbb{R}^n} \phi\, |\tilde{u}|^{p-2} \tilde{u} \partial_i \overline{\tilde{u}}$$

$$-\int_{\mathbb{R}^n} \phi\, \frac{p-2}{2} |\tilde{u}|^{p-2} (\overline{\tilde{u}} \partial_i \tilde{u} + \tilde{u} \partial_i \overline{\tilde{u}})$$

$$-\int_{\mathbb{R}^n} (\partial_i \phi)\, |\tilde{u}|^p.$$

Using a test function and the fact that $u \in W_0^{1,p}(\Omega)$, we see that $\partial_i \widetilde{u} = \widetilde{\partial_i u}$. Therefore, we can replace $\mathbb{R}^n$ by Ω and leave out the tildes in (5.11) which ends the proof.

Proposition 5.2.3. *Let $\Omega \subseteq \mathbb{R}^n$ be a Lipschitz domain and let the drift operator B_Ω be defined as in (5.8). Then $B_\Omega + \frac{\operatorname{tr} M}{p}$ is dissipative in $L^p(\Omega)$ for all $1 < p < \infty$.*

Proof. By Lemma 1.3.4, we need to show that $\operatorname{Re} \langle B_\Omega u, u^* \rangle \leq -\frac{\operatorname{tr} M}{p} \|u\|_p^p$ where

$$u^*(x) = \begin{cases} |u(x)|^{p-2}\overline{u}(x) & \text{for } u(x) \neq 0, \\ 0 & \text{for } u(x) = 0. \end{cases}$$

When Ω is unbounded, the term Mx is unbounded which causes some problems. Therefore, we need to introduce a cut-off function. Let $\varphi \in C_c^\infty(\mathbb{R}^n)$, $0 \leq \varphi \leq 1$ such that $\varphi \equiv 1$ on the unit ball B_1 and $\operatorname{supp} \varphi \subseteq B_2$. Let $\varphi_R(x) := \varphi(x/R)$. Then $|\nabla \varphi_R(x)| \leq C/R$ and $\nabla \varphi_R$ converges pointwise to 0, while φ_R converges pointwise to 1. We have

$$\begin{aligned} \operatorname{Re} \langle B_\Omega u, u^* \rangle &= \operatorname{Re} \int_\Omega Mx \cdot \nabla u \, |u|^{p-2}\overline{u} \\ &= \operatorname{Re} \sum_{i,j=1}^n \int_\Omega m_{ij}x_j(\partial_i u) \, |u|^{p-2}\overline{u} \\ &= \operatorname{Re} \sum_{i,j=1}^n \lim_{R \to \infty} \int_\Omega \varphi_R \, m_{ij}x_j(\partial_i u) \, |u|^{p-2}\overline{u}, \end{aligned}$$

where the last equality holds by Lebesgue's Dominated Convergence Theorem. Now we apply Proposition 5.2.2 with $\phi = \varphi_R m_{ij}x_j$ to obtain

$$\begin{aligned} \int_\Omega \varphi_R \, m_{ij}x_j(\partial_i u) \, |u|^{p-2}\overline{u} &= -\int_\Omega \varphi_R \, m_{ij}x_j \, |u|^{p-2}u\partial_i\overline{u} \\ &\quad -\int_\Omega \varphi_R \, m_{ij}x_j \frac{p-2}{2}|u|^{p-2}(\overline{u}\partial_i u + u\partial_i\overline{u}) \\ &\quad -\int_\Omega ((\partial_i\varphi_R) \, m_{ij}x_j + \varphi_R \, \delta_{ij}m_{ij}) \, |u|^p. \end{aligned}$$

On the left hand side and the first two terms on the right hand side, the unbounded term $m_{ij}x_j$ only appears together with $\partial_i u$. Since $Mx \cdot \nabla u \in$

$L^p(\Omega)$ and the remaining terms lie in $L^{p'}(\Omega)$ we can take the limit $R \to \infty$ into these integrals. For the last integral note that $|\nabla\varphi_R(x)| \leq C/R$ and that $|m_{ij}x_j| \leq CR$ on supp $\nabla\varphi_R$, so we can apply the Dominated Convergence Theorem here, as well. Collecting all these results, we have

$$
\begin{aligned}
\mathrm{Re}\ \langle B_\Omega u, u^* \rangle &= -\mathrm{Re} \sum_{i,j=1}^n \int_\Omega m_{ij}x_j\, |u|^{p-2}u\partial_i\overline{u} \\
&\quad -\mathrm{Re} \sum_{i,j=1}^n \int_\Omega m_{ij}x_j\, \frac{p-2}{2}|u|^{p-2}(\overline{u}\partial_i u + u\partial_i\overline{u}) \\
&\quad -\mathrm{Re} \sum_{i,j=1}^n \int_\Omega \delta_{ij}m_{ij}\, |u|^p \\
&= -(p-1)\ \mathrm{Re}\ \langle B_\Omega u, u^* \rangle - \mathrm{tr}\, M\, \|u\|_p^p.
\end{aligned}
$$

This yields

$$
\mathrm{Re}\ \langle B_\Omega u, u^* \rangle = -\frac{\mathrm{tr}\, M}{p}\, \|u\|_p^p.
$$

Therefore, the operator $B_\Omega - \lambda$ is dissipative for $\lambda \geq -\mathrm{tr}\, M/p$.   □

5.3   Ornstein-Uhlenbeck operators on bounded Lipschitz domains

Let D be a bounded Lipschitz domain in $\mathbb{R}^n$, $n \geq 3$. We study the operator

$$
\text{(5.12)} \qquad
\begin{cases}
A_D u(x) := \Delta u(x) + Mx \cdot \nabla u(x), \ x \in D, \\
D(A_D) := \{u \in W_0^{1,p}(D) : \Delta u \in L^p(D)\}.
\end{cases}
$$

By Theorem 3.3.1, the Laplacian with domain $D(A_D)$ generates an analytic C_0-semigroup on $L^p(D)$ for $(3+\varepsilon)' < p < 3+\varepsilon$ where $\varepsilon > 0$ depends only on the Lipschitz constant of D. Our main result for bounded domains is now an easy consequence of perturbation results for C_0-semigroups.

Proposition 5.3.1. *On bounded Lipschitz domains $D \subseteq \mathbb{R}^n$, the operator A_D as defined in (5.12) generates an analytic quasi-contractive semigroup T*

for $(3 + \varepsilon)' < p < 3 + \varepsilon$ where $\varepsilon > 0$ depends on the Lipschitz constant of the domain D. The growth bound of the semigroup can be estimated by

$$\omega(T) \leq -\frac{\operatorname{tr} M}{p}.$$

For $\operatorname{Re}\lambda > -\frac{\operatorname{tr} M}{p}$, there exists a constant C_λ depending on λ, such that we have the estimate

(5.13) $\quad \|u\|_{L^p(D)} + \|\Delta u\|_{L^p(D)} + \|Mx \cdot \nabla u\|_{L^p(D)} \leq C_\lambda \|f\|_{L^p(D)}$

for solutions of the resolvent problem $(\lambda - A_D)u = f$. Furthermore, the solution u satisfies the estimate

(5.14) $$\|u\|_{L^p(D)} \leq \frac{C}{|\lambda + \frac{\operatorname{tr} M}{p}|} \|f\|_{L^p(D)}$$

for $\lambda \in -\frac{\operatorname{tr} M}{p} + \Sigma_\varphi$ for some angle φ, and, for $\lambda > \max\left\{-\frac{\operatorname{tr} M}{p}, 0\right\}$, there exists $\Theta > 0$ such that

(5.15) $$\|\nabla u\|_{L^p(D)} \leq \frac{C}{\lambda^\Theta} \|f\|_{L^p(D)}.$$

Proof.   Since the weak Dirichlet-Laplacian generates an analytic semigroup of contractions by Corollary 3.1.13, Theorem 1.3.12 and Proposition 5.2.1 imply that A_D generates an analytic C_0-semigroup. The fact that it is quasi-contractive and the growth bound estimate follow from the dissipativity of the operator $B_D + \frac{\operatorname{tr} M}{p}$ (Proposition 5.2.3) and Theorem 1.3.10. Estimate (5.13) follows from the closed graph theorem, while (5.14) is the standard resolvent estimate for analytic semigroups. For (5.15), we use the representation of the resolvent as

$$R(\lambda, A_D) = R(\lambda, \Delta_D) \sum_{n=0}^{\infty} (B_D R(\lambda, \Delta_D))^n,$$

where Δ_D denotes the Dirichlet-Laplacian on D and $\lambda > \max\left\{-\frac{\operatorname{tr} M}{p}, 0\right\}$. Due to the relative boundedness, the Neumann series gives a bounded operator, and we only need to show the corresponding estimate for the Laplacian. Now, a similar calculation as in the proof of Proposition 5.2.1 shows that for $u \in D(A_D)$ we have

$$\|u\|_{1,p} \;\leq\; C\lambda^{-\Theta} \|(\lambda - \Delta)u\|_p + C\lambda^{1-\Theta} \|u\|_p.$$

Again using the fact that the Dirichlet-Laplacian generates an analytic semi-group of contractions, we can estimate $\lambda \left\| u \right\|_p \leq C \left\| (\lambda - \Delta)u \right\|_p$ to obtain

$$\left\| u \right\|_{1,p} \leq C\lambda^{-\Theta} \left\| (\lambda - \Delta)u \right\|_p.$$

Therefore,

$$\left\| \nabla R(\lambda, A_D) \right\|_{\mathcal{L}(L^p(D))} \leq C\lambda^{-\Theta}$$

concluding the proof. □

5.4 Ornstein-Uhlenbeck operators on exterior Lipschitz domains

Let $\Omega = \mathbb{R}^n \setminus \overline{K}$ where K is a bounded Lipschitz domain contained in a ball B_R of radius R. Again we define the Ornstein-Uhlenbeck operator

$$(5.16)\begin{cases} A_\Omega u(x) := \Delta u(x) + Mx \cdot \nabla u(x), \ x \in \Omega, \\ D(A_\Omega) := \{u \in W_0^{1,p}(\Omega) : \Delta u \in L^p(\Omega) \text{ and } Mx \cdot \nabla u \in L^p(\Omega)\}. \end{cases}$$

Using the Lumer-Phillips Theorem, we will see that A_Ω is a closed operator and generates a C_0-semigroup on $L^p(\Omega)$ for $(3 + \varepsilon)' < p < 3 + \varepsilon$. In a first step, our aim is to show that the range condition $(\lambda - A_\Omega)D(A_\Omega) = L^p(\Omega)$ is satisfied for some $\lambda > 0$ and $(3 + \varepsilon)' < p < 3 + \varepsilon$.

We use the following notation. The bounded Lipschitz domain $\Omega \cap B_{R+3}$ will be denoted by D. We further introduce a cut-off function $\varphi \in C^\infty(\mathbb{R}^n)$ with

$$\varphi(x) = \begin{cases} 0 & \text{for } x \in B_{R+1} \\ 1 & \text{for } x \in B_{R+2}^c. \end{cases}$$

Given $f \in L^p(\Omega)$, denote its trivial extension to K by f_0 and its restriction to D by f_*. By u_0 we denote the solution to the resolvent problem

$$(\lambda - A_{\mathbb{R}^n})u = f_0 \text{ in } \mathbb{R}^n$$

given in Theorem 5.1.1. Let u_* denote the solution to

$$(\lambda - A_D)u = f_* \text{ in } D$$

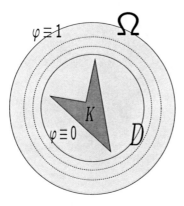

Figure 5.1: The domain D and the cut-off function φ.

given by Proposition 5.3.1. Note that

$$u_0 \in W^{2,p}(\mathbb{R}^n) \cap \{u \in L^p(\mathbb{R}^n) : Mx \cdot \nabla u \in L^p(\mathbb{R}^n)\},$$
$$u_* \in W_0^{1,p}(D) \text{ and } \Delta u_* \in L^p(D).$$

For the solution of the resolvent problem

$$(5.17) \qquad\qquad (\lambda - A_\Omega)u = f$$

we make the ansatz

$$u = \Theta_\lambda f := \varphi u_0 + (1 - \varphi)u_*.$$

The function u then satisfies

$$\Delta u = \varphi \Delta u_0 + (1 - \varphi)\Delta u_* + 2\nabla\varphi \cdot (\nabla u_0 - \nabla u_*) + \Delta\varphi(u_0 - u_*)$$

and

$$Mx \cdot \nabla u = \varphi\, Mx \cdot \nabla u_0 + (1 - \varphi)Mx \cdot \nabla u_* + (Mx \cdot \nabla\varphi)(u_0 - u_*).$$

Due to the properties of u_0 and u_*, it is easy to check that $u \in W_0^{1,p}(\Omega)$, $\Delta u \in L^p(\Omega)$ and $Mx \cdot \nabla u \in L^p(\Omega)$. Furthermore, by estimates (5.3) and (5.13), we have the estimate

$$(5.18) \quad \|u\|_{L^p(\Omega)} + \|\Delta u\|_{L^p(\Omega)} + \|Mx \cdot \nabla u\|_{L^p(\Omega)} \le C_{\lambda,\varphi} \|f\|_{L^p(\Omega)}.$$

Moreover, u satisfies the following equation.

$$\begin{aligned} \lambda u - \Delta u - Mx \cdot \nabla u &= \varphi f_0 + (1 - \varphi)f_* - 2\nabla\varphi \cdot (\nabla u_0 - \nabla u_*) \\ &\quad -\Delta\varphi(u_0 - u_*) - (Mx \cdot \nabla\varphi)(u_0 - u_*) \\ &= (I - T_\lambda)f, \end{aligned}$$

where

$$T_\lambda f := 2\nabla\varphi \cdot (\nabla u_0 - \nabla u_*) + \Delta\varphi(u_0 - u_*) + (Mx \cdot \nabla\varphi)(u_0 - u_*).$$

Our aim is now to invert $I - T_\lambda$ continuously in $\mathcal{L}(L^p(\Omega))$. To do this we need the gradient estimate (5.4) for the whole space solution u_0, and for u_* we use the estimates (5.14) and (5.15). Noting that $\nabla\varphi$ is supported in a compact set, for sufficiently large $\lambda > 0$ this yields

$$\begin{aligned} \|T_\lambda f\|_p &\le C\|\nabla u_0\|_p + C\|\nabla u_*\|_p + C\|u_0\|_p + C\|u_*\|_p \\ &\le \frac{C}{|\lambda|^\Theta}\|f\|_p + \frac{C}{|\lambda|}\|f\|_p \\ &\le \frac{1}{2}\|f\|_p. \end{aligned}$$

Therefore, for sufficiently large $\lambda > 0$, $I - T_\lambda$ can be continuously inverted in $\mathcal{L}(L^p(\Omega))$ using the Neumann series. Then we can solve the resolvent problem (5.17) by setting $u = \Theta_\lambda(I - T_\lambda)^{-1}f$.

We can now state the main result of this chapter.

Theorem 5.4.1. *Let Ω be an exterior Lipschitz domain. Then for $(3+\varepsilon)' < p < 3+\varepsilon$ where $\varepsilon > 0$ depends on the Lipschitz constant of the domain Ω, the operator A_Ω defined as in (5.16) generates a quasi-contractive C_0-semigroup T on $L^p(\Omega)$. The growth bound can be estimated by $\omega(T) \le -\frac{\operatorname{tr} M}{p}$.*

Proof. The preceding discussion shows that the range condition of the Lumer-Phillips-Theorem is satisfied. By Proposition 5.2.3, the operator $B_\Omega + \frac{\operatorname{tr} M}{p}$ is dissipative. For $p \geq 2$, the weak Dirichlet-Laplacian $\Delta_{p,w}^D$ is dissipative by Lemma 3.1.4. This proves the theorem for $2 \leq p < 3 + \varepsilon$. For $(3 + \varepsilon)' < p < 2$ we have so far only shown dissipativity of the weak Laplacian for bounded Lipschitz domains. However, simply setting $M = 0$ in the preceding discussion, we know that $\Delta_{p,w}^D$ is m-dissipative in exterior Lipschitz domains for $2 \leq p < 3 + \varepsilon$. The same consideration as in the proof of Theorem 3.1.9 then proves dissipativity of $\Delta_{p,w}^D$ and therefore of $A_\Omega + \frac{\operatorname{tr} M}{p}$ also for $(3 + \varepsilon)' < p < 2$ in exterior Lipschitz domains. $\qquad \Box$

5.5 Domains satisfying an outer ball condition

We now want to consider equation (5.1) in domains Ω satisfying a uniform outer ball condition where we can show that the domain of the operator $\Delta + Mx \cdot \nabla$ is contained in $W^{2,p}(\Omega)$. To do this we proceed in the same manner as for general Lipschitz domains using the added regularity of solutions to the heat equation we proved in Theorem 3.3.2 for bounded Lipschitz domains satisfying a uniform outer ball condition.

5.5.1 Bounded domains

Let D be a bounded Lipschitz domain satisfying a uniform outer ball condition (cf. Definition 2.1.3). On D, we define the operator

$$(5.19) \qquad \begin{cases} A_D u(x) = \Delta u(x) + Mx \cdot \nabla u(x), \ x \in D, \\ D(A_D) = W^{2,p}(D) \cap W_0^{1,p}(D). \end{cases}$$

Proposition 5.5.1. *The drift operator B_D on D given by*

$$\begin{cases} B_D u(x) = Mx \cdot \nabla u(x), \ x \in D, \\ D(B_D) = W^{1,p}(D) \end{cases}$$

is relatively bounded by the Laplacian in $L^p(D)$ for $1 < p \leq 2$. The relative bound is given by $a_0 = 0$.

Proof. For $1 < p \leq 2$, the Laplacian with domain $D(\Delta) = W^{2,p}(D) \cap W_0^{1,p}(D)$ is a closed operator in $L^p(D)$ (cf. Theorem 3.3.2). Obviously, $D(B_D) \supseteq D(\Delta)$. Furthermore, using Ehrling's Lemma (cf. Lemma 1.2.13) and Theorem 2.1.2, for $u \in D(\Delta)$, we have for any $\varepsilon > 0$,

$$\|Mx \cdot \nabla u\|_p \leq C \|\nabla u\|_p \leq C(\varepsilon \|\nabla^2 u\|_p + C(\varepsilon) \|u\|_p)$$
$$\leq \varepsilon \|\Delta u\|_p + C(\varepsilon) \|u\|_p. \qquad \square$$

As in the case of an arbitrary bounded Lipschitz domain, our generation result for bounded Lipschitz domains satisfying a uniform outer ball condition is now an easy consequence of the proposition and results from the perturbation theory of generators of contractive and analytic C_0-semigroups (Theorems 1.3.10 and 1.3.12).

Corollary 5.5.2. *For $1 < p \leq 2$, on bounded Lipschitz domains D satisfying a uniform outer ball condition, the operator A_D as defined in (5.19) generates an analytic quasi-contractive semigroup T with $\omega(T) \leq -\frac{\operatorname{tr} M}{p}$. For $\operatorname{Re}\lambda > -\frac{\operatorname{tr} M}{p}$, there exists a constant C_λ depending on λ such that we have the estimate*

$$(5.20) \qquad \|u\|_{W^{2,p}(D)} + \|Mx \cdot \nabla u\|_{L^p(D)} \leq C_\lambda \|f\|_{L^p(D)}$$

for solutions of the resolvent problem

$$(\lambda - A_D)u = f.$$

We now show that we also get so-called L^p-L^q-smoothing estimates for the generated semigroups. These kind of estimates are useful when studying semilinear equations. For example, in [HS04], L^p-L^q-smoothing estimates for the solution of the Stokes equation with linearly growing initial data are used to obtain mild solutions for the corresponding Navier-Stokes equations.

Lemma 5.5.3. *Let D be a bounded Lipschitz domain satisfying a uniform outer ball condition. Let $1 < p \leq q \leq 2$ and denote the semigroup generated* *by the operator* A_D *by* e^{tA_D}. *Then for any* $\omega > -\min\left\{\frac{\operatorname{tr} M}{p}, \frac{\operatorname{tr} M}{q}\right\}$ *there exists a constant* C *such that for all* $f \in L^p(D)$,

$$(5.21) \qquad \left\| e^{tA_D} f \right\|_q \leq C t^{-\frac{n}{2}\left(\frac{1}{p} - \frac{1}{q}\right)} e^{\omega t} \|f\|_p, \ t > 0,$$

and

$$(5.22) \qquad \left\| \nabla e^{tA_D} f \right\|_q \leq C t^{-\frac{1}{2} - \frac{n}{2}\left(\frac{1}{p} - \frac{1}{q}\right)} e^{\omega t} \|f\|_p, \ t > 0.$$

Proof. The proof is standard and relies on analyticity of the semigroup and the Gagliardo-Nirenberg inequality (Lemma 1.2.14). We include it here for completeness. We first note that the semigroups generated by A_D on the spaces $L^p(\Omega)$, $1 < p \leq 2$, are consistent which can be shown similarly as in the proof of Proposition 3.1.12 for the Laplacian. By (5.20) we have

$$\left\| \nabla^2 u \right\|_s \leq C \left\| (\lambda - A_D) u \right\|_s \leq C (\|A_D u\|_s + \|u\|_s),$$

for sufficiently large λ, any $u \in D(A_D)$ and $1 < s \leq 2$.

Now choose $q_1 \in [p, q]$ such that $\frac{1}{q_1} \leq \frac{1}{n} + \frac{1}{q}$ and $\tilde{\omega}$ such that

$$-\min\left\{\frac{\operatorname{tr} M}{p}, \frac{\operatorname{tr} M}{q}\right\} < \tilde{\omega} < \omega.$$

Then we have that $a := \frac{n}{2}\left(\frac{1}{q_1} - \frac{1}{q}\right) \leq \frac{1}{2}$. Using the Gagliardo-Nirenberg inequality (1.1) with $j = 0$, $l = 2$, we then get

$$
\begin{aligned}
\left\| e^{tA_D} f \right\|_q
&\leq C \left\| \nabla^2 e^{tA_D} f \right\|_{q_1}^a \left\| e^{tA_D} f \right\|_{q_1}^{1-a} + C \left\| e^{tA_D} f \right\|_{q_1} \\
&\leq C \left\| A_D e^{tA_D} f \right\|_{q_1}^a \left\| e^{tA_D} f \right\|_{q_1}^{1-a} + C \left\| e^{tA_D} f \right\|_{q_1} \\
&\leq C \left\| A_D e^{\frac{t}{2}A_D} e^{\frac{t}{2}A_D} f \right\|_{q_1}^a \left\| e^{\frac{t}{2}A_D} e^{\frac{t}{2}A_D} f \right\|_{q_1}^{1-a} + C \left\| e^{\frac{t}{2}A_D} e^{\frac{t}{2}A_D} f \right\|_{q_1} \\
&\leq C t^{-a} e^{\frac{a\tilde{\omega}t}{2}} \left\| e^{\frac{t}{2}A_D} f \right\|_{q_1}^a e^{\frac{(1-a)\tilde{\omega}t}{2}} \left\| e^{\frac{t}{2}A_D} f \right\|_{q_1}^{1-a} + C e^{\frac{\tilde{\omega}t}{2}} \left\| e^{\frac{t}{2}A_D} f \right\|_{q_1} \\
&\leq C t^{-\frac{n}{2}\left(\frac{1}{q_1} - \frac{1}{q}\right)} e^{\frac{\tilde{\omega}t}{2}} \left\| e^{\frac{t}{2}A_D} f \right\|_{q_1} + C e^{\frac{\tilde{\omega}t}{2}} \left\| e^{\frac{t}{2}A_D} f \right\|_{q_1} \\
&\leq C t^{-\frac{n}{2}\left(\frac{1}{q_1} - \frac{1}{q}\right)} e^{\frac{\omega t}{2}} \left\| e^{\frac{t}{2}A_D} f \right\|_{q_1}
\end{aligned}
$$

for $t > 0$ and $f \in L^p(\Omega)$. Now choose $q_2 \in [p, q_1]$ such that $\frac{1}{q_2} \leq \frac{1}{n} + \frac{1}{q_1} \leq \frac{2}{n} + \frac{1}{q}$. By a similar calculation we see that

$$\left\| e^{tA_D} f \right\|_q \leq C t^{-\frac{n}{2}\left(\frac{1}{q_2} - \frac{1}{q}\right)} e^{\frac{3\omega t}{4}} \left\| e^{\frac{t}{4}A_D} f \right\|_{q_2}$$

for $t > 0$ and $f \in L^p(\Omega)$. Iterating this procedure, we obtain q_m such that $\frac{1}{q_m} \leq \frac{m}{n} + \frac{1}{q}$ and

$$\left\| e^{tA_D} f \right\|_q \leq Ct^{-\frac{n}{2}(\frac{1}{q_m} - \frac{1}{q})} e^{\frac{(2^m-1)\omega t}{2^m}} \left\| e^{\frac{t}{2^m} A_D} f \right\|_{q_m}$$

for $t > 0$ and $f \in L^p(\Omega)$. After finitely many steps, we reach p and then (5.21) obviously is satisfied.

For (5.22), we use the Gagliardo-Nirenberg inequality in the form

$$\left\| \nabla u \right\|_q \leq C(\left\| \nabla^2 u \right\|_{q_1} + \left\| u \right\|_{q_1})^a \left\| u \right\|_{q_1}^{1-a}$$

where we choose $q_1 \in [p, q]$ such that $\frac{1}{q_1} \leq \frac{1}{2n} + \frac{1}{q}$ and $a := \frac{n}{2}(\frac{1}{q_1} - \frac{1}{q}) + \frac{1}{2} \leq \frac{3}{4}$. Proceeding similarly to above then leads to the desired gradient estimate.

$\square$

5.5.2 Exterior domains

Let Ω be an exterior Lipschitz domain in $\mathbb{R}^n$ satisfying a uniform outer ball condition. By this we mean that the complement of Ω is a compact set and that Ω itself is a Lipschitz domain and satisfies a uniform outer ball condition.

Remarks 5.5.4. Examples for exterior domains satisfying a uniform outer ball condition:

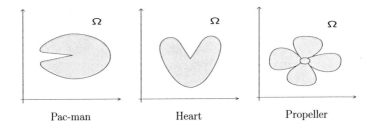

Pac-man   Heart   Propeller

Again we mimic the proof of the general case with the different domain of the operators.

Let the operator A_Ω be defined by

$$(5.23) \quad \begin{cases} A_\Omega u(x) = \Delta u(x) + Mx \cdot \nabla u(x), \ x \in \Omega, \\ D(A_\Omega) = \{u \in W^{2,p}(\Omega) : Mx \cdot \nabla u \in L^p(\Omega)\} \cap W_0^{1,p}(\Omega). \end{cases}$$

Then, as by Corollary 3.1.7, the strong Dirichlet-Laplacian is dissipative in $L^p(\Omega)$ for $1 < p \leq 2$, Proposition 5.2.3 implies that $A_\Omega + \frac{\mathrm{tr}\ M}{p}$ is dissipative in $L^p(\Omega)$ for $1 < p \leq 2$.

We now have to show that the range condition $(\lambda - A_\Omega)D(A_\Omega) = L^p(\Omega)$ is satisfied for some $\lambda > 0$ and $1 < p \leq 2$. Using the same notation and proceeding just as before, we note that now D is a bounded Lipschitz domain satisfying a uniform outer ball condition. Therefore, $u_* \in W^{2,p}(D) \cap W_0^{1,p}(D)$ and $u := \varphi u_0 + (1 - \varphi)u_* \in W^{2,p}(\Omega) \cap W_0^{1,p}(\Omega)$. Instead of (5.18), we have the stronger estimate

$$\|u\|_{W^{2,p}(\Omega)} + \|Mx \cdot \nabla u\|_{L^p(\Omega)} \leq C_{\lambda,\varphi} \|f\|_{L^p(\Omega)}.$$

We summarise the generator result of this section in the following theorem.

Theorem 5.5.5. *Let Ω be an exterior Lipschitz domain satisfying a uniform outer ball condition. Then for $1 < p \leq 2$, the operator A_Ω defined as in (5.23) generates a quasi-contractive C_0-semigroup T on $L^p(\Omega)$. The growth bound can be estimated by $\omega(T) \leq -\frac{\mathrm{tr}\ M}{p}$.*

Finally, we can extend the L^p-L^q-smoothing estimates for the generated semigroups to exterior domains. In order to do this, we use a lemma on iterated convolutions proved in [GHH05]:

Lemma 5.5.6. *Let X, Y be Banach spaces and let $T : (0, \infty) \to \mathcal{L}(Y, X)$ and $S : (0, \infty) \to \mathcal{L}(Y)$ be strongly continuous functions. Assume that*

$$\|T(t)\|_{\mathcal{L}(Y,X)} \leq C_0 t^\alpha e^{\omega t}, \quad \|S(t)\|_{\mathcal{L}(Y)} \leq C_0 t^\beta e^{\omega t}, \quad t > 0,$$

for some $C_0, \omega > 0$ and $\alpha, \beta > -1$. For $f \in Y$, set $T_0(t)f := T(t)f$ and

$$T_n(t)f := \int_0^t T_{n-1}(t - s)S(s)f\,\mathrm{d}s, \quad n \in \mathbb{N}, \ t > 0.$$

Then there exist $C, \tilde{\omega} > 0$ such that

$$\sum_{n=0}^{\infty} \|T_n(t)f\|_X \le Ct^{\alpha} e^{\tilde{\omega}t} \|f\|_Y, \quad t > 0.$$

Just as in the case of bounded domains, the statement of the L^p-L^q-smoothing properties in exterior domains reads as follows:

Theorem 5.5.7. *Let $1 < p \le q \le 2$ and let Ω be an exterior Lipschitz domain satisfying a uniform outer ball condition. Denote the semigroup generated by A_Ω by e^{tA_Ω}. Then there exist constants C, ω such that for all $f \in L^p(\Omega)$,*

$$(5.24) \qquad \left\| e^{tA_\Omega} f \right\|_q \le Ct^{-\frac{n}{2}(\frac{1}{p} - \frac{1}{q})} e^{\omega t} \|f\|_p, \; t > 0,$$

and

$$(5.25) \qquad \left\| \nabla e^{tA_\Omega} f \right\|_q \le Ct^{-\frac{1}{2} - \frac{n}{2}(\frac{1}{p} - \frac{1}{q})} e^{\omega t} \|f\|_p, \; t > 0.$$

Remarks 5.5.8. The constant ω here is not given precisely by the growth bound of the semigroup as is the case when considering bounded domains.

Proof. The proof is the same as the proof in [GHHW05] for $C^{1,1}$-domains and is included here for completeness. Recall from the construction in Section 5.4 that for sufficiently large λ, the resolvent of $\lambda - A_\Omega$ has the form

$$(\lambda - A_\Omega)^{-1} f = \Theta_\lambda (I - T_\lambda)^{-1} f, \; f \in L^p(\Omega)$$

with Θ_λ and T_λ as in Section 5.4. Let f_0 denote the trivial extension of f to $\mathbb{R}^n$ and f_D denote the restriction of f to D. Then the Laplace transforms of the strongly continuous functions $U : [0, \infty) \to \mathcal{L}(L^p(\Omega))$ and $V : [0, \infty) \to \mathcal{L}(L^p(\Omega))$ defined by

$$\begin{aligned} U(t)f &:= \varphi e^{tA_{\mathbb{R}^n}} f_0 + (1 - \varphi) e^{tA_D} f_D, \\ V(t)f &:= -2\nabla\varphi \cdot \nabla(e^{tA_{\mathbb{R}^n}} f_0 - e^{tA_D} f_D) \\ &\quad - [\Delta\varphi + (Mx \cdot \nabla\varphi)](e^{tA_{\mathbb{R}^n}} f_0 - e^{tA_D} f_D) \end{aligned}$$

are given by Θ_λ and T_λ, respectively. By the estimates (5.5), (5.7), (5.21) and (5.22), there exist constants C, ω such that

$$\|U(t)\|_{\mathcal{L}(L^p(\Omega))} \le Ce^{\omega t}, \qquad \|V(t)\|_{\mathcal{L}(L^p(\Omega))} \le Ct^{-\frac{1}{2}} e^{\omega t}, \qquad t > 0.$$

For $f \in L^p(\Omega)$ we set $T_0(t)f := U(t)f$ and define

$$T_n(t)f := \int_0^t T_{n-1}(t-s)V(s)f \, ds, \quad n \in \mathbb{N}, \quad t > 0.$$

It then follows from Lemma 5.5.6 that,

$$T_\Omega(t)f := \sum_{n=0}^\infty T_n(t)f$$

is well defined for all $t > 0$ and exponentially bounded. Thus, by Lebesgue's theorem

$$\int_0^\infty e^{-\lambda t} T_\Omega(t) dt = \sum_{n=0}^\infty \int_0^\infty e^{-\lambda t} T_n(t) dt = \sum_{n=0}^\infty \hat{U}(\lambda)\hat{V}(\lambda)^n = (\lambda - A_\Omega)^{-1}$$

for λ large enough and hence $T_\Omega(t) = e^{tA_\Omega}$ for $t \geq 0$.

Now, consider U as a mapping $U : [0, \infty) \rightarrow \mathcal{L}(L^p(\Omega), L^q(\Omega))$. Once more using the estimates (5.5), (5.7), (5.21) and (5.22), there exist constants C, ω such that

$$\|U(t)\|_{\mathcal{L}(L^p(\Omega), L^q(\Omega))} \leq Ct^{-\frac{n}{2}(\frac{1}{p} - \frac{1}{q})} e^{\omega t}, \quad t > 0.$$

Then (5.24) follows as above by Lemma 5.5.6 if $\frac{n}{2}(\frac{1}{p} - \frac{1}{q}) < 1$. In the general case, choose q_1 such that $\frac{n}{2}(\frac{1}{p} - \frac{1}{q_1}) < 1$ and proceed as above. Then repeat this procedure until q is reached.

Estimate (5.25) follows in a similar way by defining

$$W(t)f := \nabla U(t)f,$$

replacing $U(t)$ in the above proof by $W(t)$ and using the estimates (5.6) and (5.22) on the gradient of the semigroup. $\qquad \square$

Remarks 5.5.9. Note that due to the representation of the semigroup using only the semigroups on the whole space $\mathbb{R}^n$ and on the bounded domain D which are consistent, the semigroups on exterior domains are also consistent.

Bibliography

[Ada75]   R.A.Adams: *Sobolev Spaces*, Academic Press, New York, 1975.

[Ado92]   V.Adolfsson: L^2-integrability of the second order derivatives for Poisson's equation in nonsmooth domains. *Math. Scand.* **70** (1992), 146-160.

[Ado93]   V.Adolfsson: L^p-integrability of the second order derivatives of Green potentials in convex domains. *Pac. J. Math.* **159**, 2 (1993), 201-225.

[AJ94]   V.Adolfsson, D.Jerison: L^p-integrability of the second order derivatives for the Neumann problem in convex domains. *Indiana Univ. Math. J.* **43**, 4 (1994), 1123-1138.

[Agm65]   S.Agmon: *Lectures on Elliptic Boundary Value Problems.* Van Nostrand Mathematical Studies **2**, New York, 1965.

[ADN59]   S.Agmon, A.Douglis, L.Nirenberg: Estimates near the boundary for solutions of elliptic partial differential equations satisfying general boundary conditions, I. *Comm. Pure Appl. Math.* **12** (1959), 623-727. II. *Comm. Pure Appl. Math.* **17** (1964), 35-92.

[AN63]   S.Agmon, L.Nirenberg: Properties of solutions of ordinary differential equations in Banach space. *Comm. Pure and Appl. Math.* **16** (1963), 121-239.

[Alt85]   H.W.Alt: *Lineare Funktionalanalysis.* Springer, 1985.

[Ama95]   H.Amann: *Linear and Quasilinear Parabolic Problems, Vol. 1. Abstract Linear Theory.* Monographs in Mathematics **89**, Birkhäuser, Basel, 1995.

[Are94]   W.Arendt: Gaussian estimates and interpolation of the spectrum in
           L^p. *Differential Integral Equations* **7** (1994), 1153-1168.

[ABHN01]   W.Arendt, C.J.K.Batty, M.Hieber, F.Neubrander: *Vector-valued
           Laplace Transforms and Cauchy Problems.* Monographs in Mathe-
           matics **96**, Birkhäuser, Basel 2001.

[AB99]   W.Arendt, P.Bénilan: Wiener regularity and heat semigroups on
           spaces of continuous functions. *Progress in Nonlinear Diff. Eq. and
           their applications* **35** (1999), 29-49.

[Aus04]   P.Auscher: On necessary and sufficient conditions for L^p estimates of
           Riesz transforms associated to elliptic operators on $\mathbb{R}^n$ and related
           estimates. *Memoirs Amer. Math. Soc.*, to appear.

[BL76]   J.Bergh, J.Löfström: *Interpolation Spaces.* Springer, Berlin 1976.

[Bro89]   R.M.Brown: The method of layer potentials for the heat equation in
           Lipschitz cylinders. *Amer. J. Math.* **111** (1989), 339-379.

[Bro90]   R.M.Brown: The initial-Neumann Problem for the heat equation in
           Lipschitz cylinders. *Trans. AMS* **320** (1990), 1-52.

[Caf87]   L.A.Caffarelli: A Harnack inequality approach to the regularity of
           free boundaries. Part I: Lipschitz free boundaries are $C^{1,\alpha}$. *Rev. Mat.
           Iberoamericana, vol. 3, no.2* (1987), 139-162.

[Cal61]   A.P.Calderón: Lebesgue spaces of differentiable functions and distri-
           butions. *Proc. Sympos. Pure Math.* **4** (1961), 33-49.

[Cam67]   S.Campanato: Un risultato relativo ad equazioni ellittiche del secondo
           ordine di tipo non variazionale. *Ann. Scuola Norm. Pisa* **21** (1967),
           701-707.

[CH87]   P.Clément, H.J.A.M.Heijmans et al.: *One-Parameter Semigroups.*
           CWI Monographs **5**, North-Holland, Amsterdam 1987.

[Cor56]   H.O.Cordes: Über die erste Randwertaufgabe bei quasilinearen Differ-
           entialgleichungen zweiter Ordnung in mehr als zwei Variablen. *Math.
           Ann.* **131**, (1956), 278-312.

[Cow83]   M.G.Cowling: Harmonic analysis on semigroups. *Annals of Mathematics* **117** (1983), 267-283.

[CDMY96]   M.Cowling, I.Doust, A.McIntosh, A.Yagi: Banach space operators with a bounded H^∞ functional calculus. *J. Austral. Math. Soc. Ser. A* **60** (1996), no. 1, 51–89.

[Dah77]   B.E.J.Dahlberg: On estimates of harmonic measure. *Arch. Rational Mech. Anal.* **65** (1977), 272-288.

[Dah79]   B.E.J.Dahlberg: L^q-estimates for Green potentials in Lipschitz domains. *Math. Scand.* **44** (1979), 149-170.

[DK87]   B.E.J.Dahlberg, C.Kenig: Hardy spaces and the Neumann Problem in L^p for Laplace's equation in Lipschitz domains. *Ann. Math.* **125** (1987), 437-465.

[DPL95]   G.Da Prato, A.Lunardi: On the Ornstein-Uhlenbeck operator in spaces of continuous functions. *J. Func. Anal.* **131** (1995), no. 1, 94-114.

[DHP03]   R.Denk, M.Hieber, J.Prüss: $\mathcal{R}$-boundedness, Fourier multipliers and problems of elliptic and parabolic type. *Memoirs Amer. Math. Soc.*, 1-114, 2003.

[dS64]   L.de Simon: Un applicazione della teoria degli integrali singolari allo studio della equazioni differenziali lineari astratto del primo ordine. *Rend. Sem. Mat. Padova* **34** (1964), 547-558.

[DV87]   G.Dore, A.Venni: On the closedness of the sum of two closed operators. *Math. Z.* **196** (1987), 189-201.

[Dor93]   G.Dore: L^p-regularity for abstract differential equations. In: Functional Analysis and Related Topics, H.Komatsu (ed.), Lecture Notes in Math. 1540. *Springer*, 1993.

[Duo89]   X.T.Duong: H_∞ functional calculus for second order elliptic partial differential operators on L^p spaces. In: Miniconference on Operators in Analysis, I.Doust, B.Jefferies, C.Li, A.McIntosh (eds.) *Proc. Centre Math. Anal. A.N.U. Vol. 24* (1989), 91-102.

[Els96]   J.Elstrodt: *Maß- und Integrationstheorie.* Springer, 1996.

[EN00]   K.-J.Engel, R.Nagel: *One-Parameter Semigroups for Linear Evolution Equations.* Springer, 2000.

[FMM98]   E.Fabes, O.Mendez, M.Mitrea: Boundary layers on Sobolev-Besov Spaces and Poisson's equation for the Laplacian in Lipschitz domains. *J. Func. Anal.* **159** (1998), 323-368.

[FS83]   E.Fabes, S.Salsa: Estimates of caloric measure and the initial-Dirichlet problem for the heat equation in Lipschitz cylinders. *Trans. AMS* **279** (1983), 635-650.

[FS94]   R.Farwig, H.Sohr: Generalized resolvent estimates for the Stokes system in bounded and unbounded domains. *J. Math. Soc. Japan* **46, 4** (1994), 607-643.

[Fol76]   G.B.Folland: *Introduction to Partial Differential Equations.* Mathematical Notes, Princeton, 1976.

[Fra79]   L.E.Fraenkel: On regularity of the boundary in the theory of Sobolev spaces. *Proc. London Math. Soc.* (3) **39** (1979), 385-427.

[Fro93]   S.J.Fromm: Potential space estimates for Green potentials in convex domains. *Proc. Amer. Math. Soc.* **119**, 1 (1993), 225-233.

[GHH05]   M.Geissert, H.Heck, M.Hieber: L^p-theory of the Navier-Stokes flow in the exterior of a moving or rotating obstacle. *Preprint.*

[GHHW05]   M.Geissert, H.Heck, M.Hieber, I.Wood: The Ornstein-Uhlenbeck semigroup in exterior domains. *Preprint.*

[GT77]   D.Gilbarg, N.S.Trudinger: *Elliptic Partial Differential Equations of Second Order.* Springer, 1977.

[Gri99]   J.A.Griepentrog: *Zur Regularität linearer elliptischer und parabolischer Randwertprobleme mit nichtglatten Daten.* Logos Verlag, Berlin, 1999.

[Gri85]   P.Grisvard: *Elliptic Problems in Nonsmooth Domains.* Pitman, Boston, 1985.

[Gru86]   G.Grubb: *Functional Calculus of Pseudodifferential Boundary Problems.* Birkhäuser, Boston, 1986.

[Hel69]   L.L.Helms: *Introduction to Potential Theory.* Wiley, New York, 1969.

[HP97]   M.Hieber, J.Prüss: Heat kernels and maximal L^p-L^q estimates for parabolic evolution equations. *Comm. in Partial Differential Equations* **22** (1997), 1647-1669.

[HP98]   M.Hieber, J.Prüss: Functional calculi for linear operators in vector-valued L^p-spaces via the transference principle. *Adv. Differential Equations* **3** (1998), 847-872.

[HS04]   M.Hieber, O.Sawada: The Navier-Stokes equations in $\mathbb{R}^n$ with linearly growing initial data. *Preprint.*

[His99a]   T.Hishida: An existence theorem for the Navier-Stokes flow in the exterior of a rotating obstacle. *Arch. Rational Mech. Anal.* **150** (1999), 307-348.

[His99b]   T.Hishida: The Stokes operator with rotation effect in exterior domains. *Analysis* **19** (1999), 51-67.

[His00]   T.Hishida: L^2-theory for the operator $\Delta + (k \times x) \cdot \nabla$ in exterior domains. *Nihonkai Math. J.* **11** (2000), 103-135.

[JK82]   D.Jerison, C.Kenig: Boundary value problems in Lipschitz domains. In: W.Littmann (ed.): *Studies in Partial Differential Equations.* MAA Studies in Math., Vol. 23 (1982), 1-68.

[JK95]   D.Jerison, C.Kenig: The inhomogeneous Dirichlet problem in Lipschitz domains. *J. Func. Anal.* **130** (1995), 161-219.

[Kat66]   T.Kato: *Perturbation Theory for Linear Operators.* Springer, 1966.

[Kon67]   V.A.Kondratiev: Boundary problems for elliptic equations in domains with conical or angular points. *Trans. Moscow Math. Soc.* **16** (1967), 227-313.

[KMR01]   V.A.Kozlov, V.G.Maz'ya, J.Roßmann: *Spectral Problems Associated with Corner Singularities of Solutions to Elliptic Equations.* Mathematical Surveys and Monographs **85**, AMS, 2001.

[Kuf77]   A.Kufner, O.John, S.Fučik: *Function Spaces.* Noordhoff, 1977.

[LSU68]   O.Ladyzenskaja, V.A.Solonnikov, N.N.Uralceva: *Linear and Quasi-linear Equations of Parabolic Type*. Amer. Math. Soc. Transl. Math. Monographs, Providence, R.I., 1968.

[Lam87]   D.Lamberton: Equations d'évolution linéaires associeés à des semi-groupes de contraction dans les espaces L^p. *J. Func. Anal.* **71** (1987), 252-262.

[LV96]   A.Lunardi, V.Vespri: Generation of strongly continuous semigroups by elliptic operators with unbounded coefficients in $L^p(\mathbb{R}^n)$. *Rend. Instit. Mat. Univ. Trieste* **28** (1996), no. suppl., 251-279.

[MPS00]   A.Maugeri, D.Palagachev, L.Softova: *Elliptic and Parabolic Equations with Discontinuous Coefficients*. Wiley, 2000.

[Maz85]   V.G.Maz'ja: *Sobolev Spaces*. Springer, 1985.

[MP84]   V.G.Maz'ja, B.A.Plamenevskii: Estimates in L_p and Hölder classes and the Miranda-Agmon maximum principle for solutions of elliptic boundary value problems in domains with singular points on the boundary. *Amer. Math. Soc. Transl.* **123** (1984), 1-56.

[Met01]   G.Metafune: L^p-spectrum of Ornstein-Uhlenbeck operators. *Ann. Scuola Norm. Sup. Pisa Cl. Sci.* (4) **30** (2001), 97-124.

[MPRS02]   G.Metafune, J.Prüss, A.Rhandi, R.Schnaubelt: The domain of the Ornstein-Uhlenbeck operator on an L^p-space with invariant measure. *Ann. Sculoa Norm. Sup. Pisa Cl. Sci.* (5) **1** (2002), 471-485

[MPRS04]   G.Metafune, J.Prüss, A.Rhandi, R.Schnaubelt: L^p-regularity for elliptic operators with unbounded coefficients. *Preprint*.

[MS05]   G.Metafune, C.Spina: An intergration by parts formula in Sobolev spaces. *Preprint*.

[NP94]   S.A.Nazarov, B.A.Plamenevskij: *Elliptic Problems in Domains with Piecewise Smooth Boundaries*. De Gruyter, Berlin, 1994.

[Nec67]   J.Nečas: *Les Méthodes Directes en Théorie des Équations Elliptiques*. Masson, Paris, 1967.

[Ouh95]   E.-M.Ouhabaz: Gaussian estimates and holomorphy of semigroups. *Proc. Amer. Math. Soc.* **123** (1995), 1465-1474.

[Ouh04]   E.-M.Ouhabaz: *Analysis of Heat Equations on Domains.* Princeton University Press, 2004.

[Paz67]   A.Pazy: Asymptotics expansions of solutions of ordinary differential equations in Hilbert space. *Arch. Rat. Mech. and Anal.* **24** (1967), 193-218.

[Paz83]   A.Pazy: *Semigroups of Linear Operators and Applications to Partial Differential Equations.* Springer, 1983.

[PRS04]   J.Prüss, A.Rhandi, R.Schnaubelt: The domain of elliptic operators on $L^p(\mathbb{R}^d)$ with unbounded drift coefficients. *Preprint.*

[RR93]   M.Renardy, R.C.Rogers: *An Introduction to Partial Differential Equations.* Springer, 1993.

[SS99]   E.Schrohe, B.-W.Schulze: Mellin and Green symbols for boundary value problems in manifolds with edges. *Integral Equations Oper. Theory* **34**, No.3 (1999), 339-363.

[Sch91]   B.-W.Schulze: *Pseudo-Differential Operators on Manifolds with Singularities.* North Holland, Amsterdam, 1991.

[She95]   Z.Shen: Resolvent estimates in L^p for elliptic systems in Lipschitz domains. *J. Func. Anal.* **133**, (1995), 224-251.

[Sob64]   P.E.Sobolevskii: Coerciveness inequalities for abstract parabolic equations. *Soviet. Math. (Doklady)* **5** (1964), 894-897.

[Sta65]   G.Stampacchia: Le problème de Dirichlet pour les équations elliptiques du second ordre à coefficients discontinus. *Ann. Inst. Fourier* **15**, 1 (1965), 189-258.

[Ste70]   E.M.Stein: *Singular Integrals and Differentiability Properties of Functions.* Princeton University Press, Princeton 1970.

[Ste93]   E.M.Stein: *Harmonic Analysis: Real-Variables Methods, Orthogonality and Oscillatory Integrals.* Princeton University Press, Princeton, 1993.

[Tal65]   G.Talenti: Sopra una classe di equazioni ellittichi a coefficienti mis-
           urabili. *Ann. Mat. Pura Appl.* **69** (1965), 285-304.

[Tay67]   A.E.Taylor: *Introduction to Functional Analysis.* Wiley, New York,
           1967.

[Tri72]   H.Triebel: *Höhere Analysis.* VEB, 1972.

[Tri78]   H.Triebel: *Interpolation Theory, Function Spaces and Differential Op-
           erators.* North-Holland, 1978.

[Tri83]   H.Triebel: *Theory of Function Spaces.* Birkhäuser, Basel, 1983.

[Wei95]   L.Weis: The stability of positive semigroups on L_p spaces. *Proc. Amer.
           Math. Soc.* **123** (1995), 3089-3094.

[Wei98]   L.Weis: A short proof for the stability theorem for positive semigroups
           on $L_p(\mu)$. *Proc. Amer. Math. Soc.* **126** (1998), 3253-3256.

[Wei01a]   L.Weis: Operator-valued Fourier multiplier theorems and maximal
           L^p-regularity. *Math.Ann.* **319** (2001), 735-758.

[Wei01b]   L.Weis: A new approach to maximal L^p-regularity. In: Evolution
           Equations and Appl. Physical Life Sciences, G.Lumer, L.Weis (eds.),
           Lect. Notes in Pure & Appl. Math., Vol.215 *Marcel Dekker, New York*
           (2001), 195-214.

[Zie89]   W.P.Ziemer: *Weakly Differentiable Functions.* Springer, 1989.